知识生产的原创基地
BASE FOR ORIGINAL CREATIVE CONTENT

颉腾科技
JIE TENG TECHNOLOGY

Advance Core Python Programming
Begin your Journey
to Master the World of Python

高级 Python 核心编程

开启精通Python编程
世界之旅

[印] 米努·科利 / 著 刘春明 曹创华 王贵财 / 译

北京理工大学出版社
BEIJING INSTITUTE OF TECHNOLOGY PRESS

版权专有　侵权必究

图书在版编目（CIP）数据

高级 Python 核心编程：开启精通 Python 编程世界之旅／（印）米努·科利著；刘春明，曹创华，王贵财译. -- 北京：北京理工大学出版社，2023.4

书名原文：Advance Core Python Programming：Begin your Journey to Master the World of Python

ISBN 978－7－5763－2226－2

Ⅰ.①高… Ⅱ.①米… ②刘… ③曹… ④王… Ⅲ.①软件工具-程序设计 Ⅳ.①TP311.561

中国国家版本馆 CIP 数据核字（2023）第 056060 号

北京市版权局著作权合同登记号　图字：01－2023－1320 号

Title：Advance Core Python Programming，by Meenu Kohli

Copyright © 2021 BPB Publications India. All rights reserved.

First published in the English language under the title Advance Core Python Programming 9789390684069 by BPB Publications India.（sales@bpbonline.com）

Chinese translation rights arranged with BPB Publications India through Media Solutions，Tokyo Japan

Simplified Chinese edition copyright © 2023 by Beijing Jie Teng Culture Media Co.，Ltd.

All rights reserved. Unauthorized duplication or distribution of this work constitutes copyright infringement.

出版发行 /	北京理工大学出版社有限责任公司
社　　址 /	北京市海淀区中关村南大街 5 号
邮　　编 /	100081
电　　话 /	（010）68914775（总编室）
	（010）82562903（教材售后服务热线）
	（010）68944723（其他图书服务热线）
网　　址 /	http：//www.bitpress.com.cn
经　　销 /	全国各地新华书店
印　　刷 /	文畅阁印刷有限公司
开　　本 /	787 毫米×1092 毫米　1/16
印　　张 / 29.5	责任编辑 / 钟　博
字　　数 / 662 千字	文案编辑 / 钟　博
版　　次 / 2023 年 4 月第 1 版　2023 年 4 月第 1 次印刷	责任校对 / 刘亚男
定　　价 / 129.00 元	责任印制 / 施胜娟

图书出现印装质量问题，请拨打售后服务热线，本社负责调换

译者序
Foreword

Python 作为一种不受局限、跨平台的开源编程语言，广泛应用于数据分析、人工智能、网络爬虫、运维测试等多个领域。利用 Python 进行项目开发与科研实践不失为明智之举。

本书适用于具备一定 Python 基础并希望通过在项目中应用最佳实践和新的开发技术来提升自己的 Python 开发人员。本书力求帮助 Python 开发人员挖掘和掌握这门语言及相关程序库的高级特性，涵盖函数、数据结构、Python 风格的对象、Python 与数据库交互、多线程、异常、数据分析及可视化、图形用户界面开发，以及 Web 应用创建等内容，并帮助读者解决常见编程问题和困惑。本书特点如下。

（1）内容全面，涵盖高级 Python 的核心技术的方方面面。

（2）由浅入深讲解实例，帮助读者逐步掌握高级 Python 的核心编程方法。

（3）图文并茂，让读者及时了解每一步的操作结果，帮助读者更好地检验学习进度。

本书将理论与实践结合，通过示例，深入浅出地介绍高级 Python 核心技术要点，并引导读者利用所学知识解决问题；通过示例和课后练习帮助读者巩固所学内容，使读者能够真正理解并应用所学内容。

本书作者和审稿人致力于 Python 应用领域，为世界各地的公司和组织建立过数据分析项目，积累了十多年的实践经验。在翻译过程中我们为作者对高级 Python 核心编程技术的深入掌握和独到见解而赞叹。翻译本书对我们而言也是一个学习与提高的过程。为了做到专业词汇权威准确，内容忠实原书，译者查阅了大量资料，但由于水平有限，加上时间仓促，错误和疏漏在所难免，恳请读者及时指出，以便再版时予以更正。我们为能向国内广大读者推荐这本好书而高兴。本书翻译分工如下：中南大学地球科学与信息物理学院刘春明负责第 1~5 章和附录，河南工业大学人工智能与大数据学院王贵财负责第 6~10 章，湖南省地质调查所曹创华负责第 11~14 章。

本书的翻译工作得到 2021 年度河南省高等教育教学改革研究与实践项目"基于产教融合的人工智能现代产业学院建设路径研究与实践"（2021SJGLX401）；2021 年度河南省高等教育教学改革研究与实践项目"地方本科高校产教融合推进高质量人才培养研究与实践"（2021SJGLX135）；2021 年河南省高等教育研究重点项目"产教融合视域下产业人才培养模式研究与实践——以人工智能为例"（2021SXHLX013）；2020 年教育部产学合作协同育人项目"面向新工科的人工智能专业课程体系建设与改革"（202002163007）；湖南省自然资源科技项目"城市体检中污染场地探测评估关键技术与应用"（2022-25）；湖南省地质院科研项目"湖南骑田岭地区深部电磁结构建模应用及找矿方向研究"（HNGSTP202201）的资助。

特别感谢参与本书资料整理的河南工业大学人工智能与大数据学院王宏伟、王子山与刘洁三位同学。特别感谢北京颉腾文化传媒有限公司李华君总经理和鲁秀敏编辑的讨论与帮助，他们的辛勤工作提高了本译著的质量。感谢家人对我们的支持与鼓励。

刘春明　曹创华　王贵财
2023 年 3 月

前言
Preface

Python 从其他编程语言中脱颖而出有很多原因。事实上,只要你现在正手握一本关于高级 Python 核心编程的书,就表明你已认识到 Python 具有如此特别的一些特性。

在本书中,你将获得有关 Python 编程高级主题的详细信息。因为本书从函数章节开始,所以,建议在阅读本书前,先温习一下 Python 的基本概念知识。如果你是初学者,建议先阅读《Python 核心编程:从入门到实践(学与练)》,以便理解本书主题。

本书的编写主要关注顶尖大学的教学内容和目前需求,还关注学生和专业人士在学习编程时面临的问题。通过与编程爱好者交流,发现许多人,因为忙于其他事情,或者只是无力承担课程费用而偏爱自学。

本书中程序开发的目的是为自学课程提供详细步骤,并且将复杂问题分解为简单问题,可以简单编码后将其重组。相信通过本书,你将能够思考、开发和创建 Python 应用程序。

本书分为 14 章:

第 1 章 函数与递归;
第 2 章 类、对象与继承;
第 3 章 文件;
第 4 章 MySQL 与 Python 交互;
第 5 章 Python 线程;
第 6 章 错误、异常、测试与调试;
第 7 章 数据可视化与数据分析 ;
第 8 章 创建 GUI 表和添加控件;
第 9 章 MySQL 和 Python 图形用户界面;
第 10 章 栈、队列和双端队列;
第 11 章 链表;
第 12 章 树;
第 13 章 查找与排序;
第 14 章 Flask 框架入门。

希望你像我热衷于编写此书一样喜欢阅读本书。
愿读书之乐乐陶陶,起并明月霜天高!

目录
Contents

第1章 函数与递归

1.1 函数//003
1.2 创建函数//003
1.2.1 简单函数//003
1.2.2 定义接收参数的函数//005
1.3 函数参数//007
1.3.1 位置参数//007
1.3.2 默认参数//008
1.3.3 关键字参数//009
1.3.4 *args//010
1.3.5 **kwargs//010
1.4 返回语句//011
1.5 作用域和命名空间//013
1.5.1 内置命名空间//014
1.5.2 全局命名空间//014
1.5.3 局部命名空间//015
1.6 lambda函数//016
1.7 递归//022
1.7.1 递归求数的阶乘//022
1.7.2 递归求阶乘的算法//023
1.7.3 递归的类型//024
1.7.4 递归的优、缺点//025
1.8 存储//027
知识要点//029
小结//030
简单题//030
编程题//031
论述题//044

第 2 章 类、对象与继承

2.1 类和对象//049
2.2 析构函数__del__( )//055
2.3 类变量的类型//056
2.4 继承//059
知识要点//063
小结//064
选择题//064
简答题//065
编程题//067
论述题//069

第 3 章 文件

3.1 文件存储的优点//075
3.2 目录与文件管理//075
3.3 使用 getcwd( )方法读取当前工作目录//076
3.4 使用 listdir( )方法读取目录内容//076
3.5 使用 mkdir( )方法创建目录//076
3.6 使用 rename( )方法重命名目录//077
3.7 使用 rmdir( )方法删除目录//078
3.8 文件操作//080
3.8.1 open( )函数与 close( )函数//080
3.8.2 文件写入的访问模式//082
3.8.3 使用 wb 访问模式写入二进制文件//086
3.8.4 使用 rb 访问模式读取二进制文件内容//087
3.8.5 使用二进制文件读写字符串//087
3.8.6 其他二进制文件操作//087
3.8.7 文件属性//088
3.9 各种操作文件的方法//088
小结//089
选择题//090
简答题//090
填空题//091
编程题//092
问答题//093

第 4 章 MySQL 与 Python 交互

4.1 安装和配置 MySQL//096
4.1.1 安装 MySQL//096
4.1.2 配置 MySQL//097
4.2 使用命令行工具在 MySQL 中创建数据库//098
4.3 连接 MySQL 数据库与 Python//099
4.3.1 使用 Python 在 MySQL 中创建数据库//099
4.3.2 使用 Python 检索数据库记录//103
4.4 创建数据库//104
4.4.1 使用 MySQL 创建数据库//104
4.4.2 使用 Python 创建数据库//106
4.5 使用 Python 操作数据库//109
4.5.1 使用 INSERT 语句插入记录//109
4.5.2 读取记录//113
4.5.3 使用 WHERE 子句读取表中选定记录//116
4.5.4 使用 ORDER BY 子句对查询结果进行排序//118
4.5.5 使用 DELETE 指令删除记录//119
4.5.6 使用 UPDATE 指令更新记录//119
小结//120
编程简答题//120

第 5 章 Python 线程

5.1 进程和线程//127
5.2 创建线程//128
5.3 使用 Lock 和 RLock 实现线程同步//140
5.4 Lock 的用法//142
5.5 死锁//144
5.5.1 使用 locked( ) 方法检查资源锁定状态//147
5.5.2 RLock 方法//148
5.6 信号量//150
5.7 使用事件对象同步线程//154
5.8 条件类//157
5.9 后台线程和非后台线程//159
小结//163
简答题//163

第 6 章 错误、异常、测试与调试

6.1 错误//172
6.1.1 语法错误//172
6.1.2 运行错误//174
6.1.3 逻辑错误//176
6.2 异常//177
6.2.1 try 和 catch//177
6.2.2 捕获通用异常//179
6.2.3 try…except…else 语句//182
6.2.4 try…except…finally 语句//183
6.2.5 try 和 finally//184
6.2.6 引发异常//184
6.3 调试程序//185
6.4 Python 调试器//189
6.5 命令行调试器//190
6.6 Python 的单元测试和测试驱动开发//191
6.7 测试级别//191
6.8 pytest 概述//192
6.9 unittest 模块//195
6.10 使用 unittest 和 pytest 定义多个测试用例//197
6.11 unittest 模块中的主要 Assert 方法//199
小结//200
简答题//200

第 7 章 数据可视化与数据分析

7.1 数据可视化//207
7.2 Matplotlib//207
7.2.1 Pyplot//208
7.2.2 绘制点//209
7.2.3 绘制多点//213
7.2.4 绘制线//214
7.2.5 标注 x 轴和 y 轴//219
7.3 Numpy//224
7.3.1 安装 Numpy//224
7.3.2 Numpy 数组形状//224
7.3.3 读取数组元素值//225
7.3.4 创建 Numpy 数组//225
7.4 Pandas//237
7.5 DataFrame 操作//243
小结//254

第 8 章 创建 GUI 表和添加控件

8.1 开始//257
8.2 控件//261
8.3 按钮和消息框//266
8.4 Canvas//274
8.5 Frame//293
8.6 标签//294
8.7 小项目——秒表//295
8.8 列表框//297
8.9 菜单按钮和菜单//299
8.10 单选按钮//307
8.11 滚动条和滑块//309
8.12 文本框//313
8.13 Spinbox//315
知识要点//315
简答题//316

第 9 章 MySQL 和 Python 图形用户界面

9.1 MySQLdb 数据库//318
9.2 使用 GUI 创建表//320
9.3 使用 GUI 插入数据//325
9.4 创建 GUI 以检索结果//328
小结//341

第 10 章 栈、队列和双端队列

10.1 栈//343
10.2 队列//351
10.2.1 基本队列函数//351
10.2.2 实现队列//352
10.2.3 使用单队列实现栈//354
10.2.4 使用两个栈实现队列//358
10.3 双端队列//361

第 11 章 链表

- 11.1 链表简介 //364
- 11.2 实现节点类 //365
  - 11.2.1 遍历链表 //367
  - 11.2.2 在链表头添加节点 //369
  - 11.2.3 在链表尾添加节点 //370
  - 11.2.4 在两个节点间插入节点 //372
  - 11.2.5 从链表中删除节点 //373
  - 11.2.6 打印链表的中心节点值 //375
  - 11.2.7 实现双向链表 //378
  - 11.2.8 反向链表 //381
- 小结 //384

第 12 章 树

- 12.1 引言 //386
- 12.2 简单树表示法 //388
- 12.3 树的列表表示 //392
- 12.4 二叉堆 //401
- 小结 //413

第 13 章 查找与排序

- 13.1 顺序查找 //415
- 13.2 对半查找 //419
- 13.3 哈希排序 //421
- 13.4 冒泡排序 //431
- 13.5 选择排序 //433
- 13.6 插入排序 //434
- 13.7 希尔排序 //437
- 13.8 快速排序 //440
- 小结 //444

第 14 章 Flask 框架入门

- 14.1 引言 //446
- 14.2 安装虚拟环境 //447
- 14.3 使用 Flask 开发 "Hello World" 应用 //451
- 14.4 调试 Flask 应用程序 //454
- 小结 //455

附录

第 1 章

函数与递归

引言 为了解 Python 基础知识、数据类型和控制结构，本章介绍如何创建函数。函数是为执行某项任务可重复调用的代码块。理解了函数概念，就会明白递归——函数调用本身解决问题。

知识结构

- 函数
- 创建函数
  - 简单函数
  - 定义接收参数的函数
- 函数参数
  - 位置参数
  - 默认参数
  - 关键字参数
  - *args
  - **kwargs
- 返回语句
- 作用域和命名空间
  - 内置命名空间
  - 全局命名空间
  - 局部命名空间
- lambda 函数
- 递归
  - 递归求数的阶乘
  - 递归求阶乘的算法
  - 递归的类型
  - 递归的优缺点
- 存储

完成本章的学习后，读者应掌握以下技能。
(1) 创建函数。
(2) 使用 lambda 函数。
(3) 递归和存储。

1.1 函数

本章中学习的函数是所有面向对象编程语言的核心主题，也就是 Python 编程的核心。结合姊妹书《Python 核心编程：从入门到实践（学与练)》的知识概念创建可调用函数。函数是可重复调用的代码块，其重要性如下。
(1) 提升可读性和模块性。
(2) 设计和执行代码有助于节省时间和精力。
(3) 减少代码重复。
(4) 代码可重复调用。
(5) 易于维护代码。
(6) 有助于理解代码的运行机制。
(7) 便于信息隐藏。

注：将程序或代码划分为独立模块的行为称为模块化。

在 Python 编程中，函数分为以下两类。
(1) 内置函数：Python 提供内置函数，可随时调用。到目前为止，所使用的函数均为内置函数，如 max( )、min( )、len( ) 函数等。
(2) 用户自定义函数：本章重点介绍用户自定义函数。这些函数并非由 Python 提供，而是因程序员为执行特定任务自行创建产生。

1.2 创建函数

函数是可重复调用的代码块。如果代码块多次运行，就可将其定义为函数，在需要时调用该函数。缩写 DRY 表示"请勿重复自身"。函数有助于保持代码不重复。DRY 与另一个缩写为 WET 的编码，其含义为"重复编写所有的内容"。

1.2.1 简单函数

本小节主要学习创建函数。
创建函数需遵守以下几点原则。

1. 函数定义

(1) 函数以 def 关键字开头。

(2) def 后面是函数名。
(3) 函数名后面紧跟括号"()"。
(4) 括号后面是冒号":",标志代码块的开始。

示例代码如下所示。

```
def helloWorldFunc():
```

2. 函数体

函数代码块向右缩进 1 级,即按照 PEP-8 规定为 4 个空格。对此,读者在学习 Python 基础知识(见《Python 核心编程:从入门到实践(学与练)》第 2 章)时便已了解。使用 if…else 语句和 for、while 等循环语句时,也有相同规则。示例代码如下所示。

```
def hello_world_func():
    '''
    This is my first function.
    I am learning a lot and this looks like fun
    This function just prints a message.
    '''
print('My first function prints HELLO WORLD!!')
```

3. 调用函数

根据函数名和括号调用函数。示例代码如下所示。

```
hello_world_func()
```

只要显式调用该函数,该函数就不会执行。

示例代码如下。

```
def hello_world_func():
    '''
    This is my first function.
    I am learning a lot and this looks like fun
    This function just prints a message.
    '''
print('My first function prints HELLO WORLD!!')
hello_world_func()
```

三引号中的字符串用于多行注释,直接放在函数、模块或类定义下充当 docstring(文档字符串)。为了便于理解,三引号中的多行注释以粗体突出显示。非强制使用 docstring,但它有助于理解复杂函数。

该函数只有单条打印语句。以下内容非强制,却是可遵循的最佳原则。

(1) 函数名使用小写字母,单词可以用下划线分隔,或采用驼峰命名法。
(2) 将 docstring 作为函数第一部分。docstring 必须强调函数功能,而不是它如何实现。
(3) 将代码放在 docstring 后。

以上代码输出结果如下所示。

```
My first function prints HELLO WORLD!!
>>>
```

在 Python shell 中输入如下命令。

```
help(hello_world_func())
```

此命令将提供有关此函数结构的详细信息,如下所示。

```
>>> help(hello_world_func())
My first function prints HELLO WORLD!!
Help on NoneType object:

class NoneType(object)
 |  Methods defined here:
 |
 |  __bool__(self, /)
 |      self != 0
 |
 |  __repr__(self, /)
 |      Return repr(self).
 |
 |  ----------------------------------------------------------------
 |
 |  Static methods defined here:
 |
 |  __new__(*args, **kwargs) from builtins.type
 |      Create and return a new object.See help(type) for accurate signature.
```

输入如下命令查看 docstring。

```
>>> print(hello_world_func.__doc__)

This is my first function.
I am learning a lot and this looks like fun
This function just prints a message.
>>>
```

1.2.2 定义接收参数的函数

1.2.1 小节中定义了简单函数。本小节修改这个简单函数,即接收一个参数(姓名),并打印单条指定消息(姓名和消息)。示例代码如下所示。

```
def hello_world_func(name):
    '''
    This is my first function.
    I am learning a lot and this looks like fun
    This function just prints a message.
    '''
    print('My first function prints HELLO WORLD!! I am {}'.format(name))
hello_world_func('Meenu')
```

注意：函数定义的括号中有一个参数名（形参）。因此，调用函数 hello_world_func( ) 时需要给出一个可传递给函数的参数（实参），如图1.1所示。

上述代码输出结果如下所示。

```
My first function prints HELLO WORLD!! I am Meenu
>>>
```

说明：

理解形参和实参之间的差异对理解代码的运行过程非常重要。

图1.1

形参是函数局部变量定义的一部分，实参是传递给函数的值。因此，name 是函数 hello_world_func( )的形参，Meenu 是调用函数 hello_world_func( )时传递给它的实参。在调用函数时，将形参 name 与实参 Meenu 映射，并使用实参值进行计算。

函数可以有多个参数，这时参数的传递顺序很重要。查看下面的代码。

```
def hello_world_func(name,age,profession):
    '''
    This is my first function.
    I am learning a lot and this looks like fun
    This function just prints a message.
    '''
    print('I am {} years old.'.format(age))
    print('My name is {}.'.format(name))
    print('I am {} by profession.'.format(profession))

hello_world_func(32,'Architect','Michael')
```

函数 hello_world_func( )依次接收3个形参 name、age 和 profession，而传递给函数的实参顺序为32、Architect、Michael。

结果，name 映射到值32，age 映射到 Architect，profession 映射到 Michael，所以运行代码时，会有如下输出结果。

```
I am Architect years old.
My name is 32.
I am Michael by profession.
```

调用函数 hello_world_func( )并以正确顺序传递参数。

代码：

```
hello_world_func('Michael',32,'Architect')
```

输出结果如下所示。

```
I am 32 years old.
My name is Michael.
I am Architect by profession.
```

Python 程序员经常交替使用实参和形参，虽然两者某种程度上非常相似，但 Python 开发人员必须理解两者的区别。形参在函数中声明，实参是调用函数时传递给函数的值。

1.3 函数参数

Python 包括 5 种函数参数。
(1) 位置参数；
(2) 默认参数；
(3) 关键字参数；
(4) *args；
(5) **kwargs。

1.3.1 位置参数

到目前为止，所有示例均使用位置参数，即按照参数传递顺序或位置将实参分配给形参。

代码：

```
def sum_prod(num1,num2):
    num_sum = num1 + num2
    num_prod = num1 * num2
    return num_sum,num_prod

x = int(input('Enter the first number :'))
y = int(input('Enter the second number :'))
print(sum_prod(x,y))
```

输出：

位置参数需要查看参数的指定位置。因此，x 映射到 num1，y 映射到 num2，并将这些值传递到函数。

如果使用不同个数的参数调用函数，将生成错误。

```
Enter the first number :10
Enter the second number :20
(30,200)
>>>
```

1.3.2 默认参数

定义函数时指定参数默认值。如果定义参数默认值，位置参数就会变为可选，因此称为默认参数。

代码：

```
def sum_prod(num1,num2 =0):
    num_sum = num1 + num2
    num_prod = num1 * num2
    return num_sum,num_prod

print(sum_prod(2,5))
print(sum_prod(2))
```

输出：

上面显示的函数可用 1 个或 2 个参数调用。如果省略第 2 个参数，函数将传递默认值 0。

再看一个示例。

```
(7,10)
(2,0)
>>>
```

Python 中的非默认参数不能紧跟其后。

```
def sum_func(num1,num2 =0,num3):
    return num1 + num2 + num3
```

其中，num3 是非默认参数，num2 是默认参数，这导致上面的函数抛出错误。如果输入 sum_func(10,20)，解释器无法理解是将 20 分配给 num2，还是继续使用默认值。随着默认参数个数的增加，复杂性也会增加。这时就会收到语法错误："non default argument follow default argument"。使用默认参数的正确方法如下。

```
def sum_func(num1,num2 =30,num3 =40):
    return num1 + num2 + num3

print(sum_func(10))
print(sum_func(10,20))
print(sum_func(10,20,30))
```

输出：

```
80
70
60
>>>
```

1.3.3 关键字参数

关键字参数允许忽略输入参数的顺序，甚至在调用函数时跳过它们。

带有关键字参数的函数定义方式与带有位置参数的函数相同，但不同之处在于调用方式。

代码：

```
def sum_func(num1,num2 =30,num3 =40):
    print("num1 = ", num1)
    print("num2 = ", num2)
    print("num3 = ", num3)
    return num1 + num2 + num3

print(sum_func(num3 = 10, num1 =20))
```

输出：

```
num1 = 20
num2 = 30
num3 = 10
60
>>>
```

如上所示，参数不是按所需顺序传递的，而是在传递参数时指定哪个参数属于哪个参数。由于 num2 的默认值为 0，所以它被跳过也没关系。这里想使用 num2 的默认值，因此只指定了 num1 和 num3 的值。如果不这样做，则输出将不正确。就像在代码中看到的，将值 20 分配给 num2，并将默认值 num3 作为结果，输出完全不同。

代码：

```
def sum_func(num1,num2 =30,num3 =40):
    print("num1 = ", num1)
    print("num2 = ", num2)
    print("num3 = ", num3)
    return num1 + num2 + num3

print(sum_func(10, 20))
```

输出：

```
num1 = 10
num2 = 20
num3 = 40
70
>>>
```

1.3.4 *args

在不知道参数的个数时，可以使用*args。如同去购物，但不知道所购买的商品数量或花费。在创建函数时，若无法定义参数个数，可以使用*args，它是有潜在附加参数的元组。该元组初始值为空，即使不提供参数，也不会产生错误。

代码：

```
def sum_func(a, *args):
    s = a + sum(args)
    print(s)
sum_func(10)
sum_func(10,20)
sum_func(10,20,30)
sum_func(10, 20, 30, 40)
```

输出：

```
10
30
60
100
>>>
```

1.3.5 **kwargs

**kwargs 表示关键字参数（可变长度），在不确定关键字参数的个数时使用。

kwargs 构建键-值对字典，在处理不同外部模块和库时，经常使用该类型参数。kwargs 中的** 允许传递任意个数的关键字参数。顾名思义，在关键字参数中，向变量提供名称，同时将其传递给类似字典的函数，字典中的关键字与值关联。

代码：

```
def shopping(**kwargs):
    print(kwargs)
    if kwargs:
        print('you bought', kwargs['dress'])
        print('you bought', kwargs['food'])
        print('you bought', kwargs['Shampoo'])
shopping(dress = 'Frock',Shampoo ='Dove',food = 'Pedigree Puppy')
```

输出：

```
{'dress':'Frock','Shampoo':'Dove','food':'Pedigree Puppy'}
you bought Frock
you bought Pedigree Puppy
you bought Dove
```

特别注意，**kwargs 类似字典，如果对其进行迭代，有可能不以相同顺序打印。

就名称而言，推荐使用 args 或 kwargs 名称，但非强制，也可以使用其他名称代替。重点要使用 * 作为位置参数，使用 ** 作为关键字参数。

1.4 返回语句

在函数结果返回调用者时，函数末尾要使用 return 关键字。进行软件编程时不一定总是打印结果。有些计算要求用户隐藏在后台执行。这些值用于计算以获得用户期望的最终输出。return 语句读取值并将其分配给变量做进一步计算。

查看以下代码。函数 adding_numbers() 将 3 个数相加的结果返回并分配给变量 x，然后输出 x 的值。

代码：

```
def adding_numbers(num1, num2, num3):
    print('Have to add three numbers')
    print('First Number = {}'.format(num1))
    print('Second Number = {}'.format(num2))
    print('Third Number = {}'.format(num3))
    return num1 + num2 + num3

x = adding_numbers(10, 20, 30)
print('The function returned a value of {}'.format(x))
```

输出：

```
Have to add three numbers
First Number = 10
Second Number = 20
Third Number = 30
The function returned a value of 60
```

总结如下。

(1) 返回语句会退出函数。返回语句是函数的最后一条语句，后面所有语句都不执行。

(2) 函数无返回值意味着可能隐式返回 None 值。

(3) 如果函数必须返回多个值，那么所有值作为元组返回。

示例 1.1

编写函数，提示用户为列表输入值。该函数返回列表长度。

代码：

```
def find_len(list1):
    return len(list1)

x = input('Enter the values separated by single space :')
x_list = x.split()
print(find_len(x_list))
```

输出：

```
Enter the values separated by single space :1 'Gmail' 'Google' 1.09
[2,3,45,9]
5
>>>
```

说明：在执行此代码程序时，将提示输入以单个空格分隔的值。这种情况下传递5个值：1、'Gmail'、'Google'、1.09、[2,3,45,9]。

因此，程序计算输入元素的个数，并且输出结果5。

示例1.2

下面的代码返回两个数的和与积的值。

代码：

```
def sum_prod(num1,num2):
    num_sum = num1 + num2
    num_prod = num1 * num2
    return num_sum,num_prod

x = int(input('Enter the first number :'))
y = int(input('Enter the second number :'))
print(sum_prod(x,y))
```

输出：

```
Enter the first number :10
Enter the second number :20
(30,200)
```

示例1.3

编写代码查找两个给定数的HCF。

HCF代表两个数的最大公因数或最大公约数，即两个给定数中1到较小范围内的最大数，并能将这两个数整除，余数为0。

(1) 定义函数hcf()，并以两个数作为输入。

```
def hcf(x,y):
```

(2) 找出这两个数中哪个最大、哪个最小。

```
small_num = 0
if x > y:
    small_num = y
else:
    small_num = x
```

设置 for 循环的范围为 1 到 small_num + 1（上限取为 small_num + 1，因为 for 循环对小于范围上限的数运算）。在 for 循环中，将两个数与范围内的每个数相除，如果有数将两者相除，则将该值赋给 hcf，代码如下所示。

```
for i in range(1,small_num +1):
    if (x % i == 0) and (y % i == 0):
        hcf = i
```

假设这两数是 6 和 24，首先都可被 2 整除，因此，hcf = 2。接着也都可被 3 整除，因此，hcf = 3。继续，循环到 6，再次将两个数整除，因此，hcf = 6。由于已达到循环范围上限，因此函数返回的 hcf 值最后为 6。

（3）返回 hcf 的值：return hcf。

代码：

```
def hcf(x,y):
    small_num = 0
    if x > y:
        small_num = y
    else:
        small_num = x
    for i in range(1,small_num +1):
        if (x % i == 0) and (y % i == 0):
            hcf = i
    return hcf
print(hcf(6,24))
```

输出：

6

1.5 作用域和命名空间

命名空间是容器，其中包含定义的所有名称（变量/函数/类）。可以在不同命名空间中定义相同名称。名称或变量存在于定义其范围代码的特定区域中。变量/对象之间绑定的信息存储在命名空间中。有 3 种类型的命名空间或作用域。

（1）内置命名空间：内置命名空间中的内置函数可用于所有应用程序文件或模块。

（2）全局命名空间：全局命名空间包含所有变量、函数，以及单个文件中的可用类。

（3）局部命名空间：局部命名空间是函数中定义的变量。

作用域嵌套，意味着全局命名空间中的局部命名空间嵌套在内置命名空间中。每个作用域都有其命名空间。

1.5.1 内置命名空间

内置命名空间适用于所有文件和 Python 模块。下面的 print( )、tuple( )、type( ) 等函数都是内置函数，都属于内置命名空间，Python 中所有文件和模块都适用。示例代码如下所示。

```
>>> list1 = [1,2,3,4,5,6]
>>> tup1 = tuple(list1)
>>> type(tup1)
<class 'tuple'>

>>> print("Hi")
Hi
>>>
```

1.5.2 全局命名空间

查看以下示例代码。

代码：

```
x = 20
x += y
print(x)
```

执行此代码时生成 NameError。

输出：

```
Traceback (most recent call last):
  File "F:\2020 - BPB\input.py", line 2, in <module>
    x += y
NameError: name 'y' is not defined
>>>
```

这是因为 Python 在全局命名空间中查找 y，但找不到，接着在内置命名空间中也找不到，因此，编译产生错误。以下代码正常运行，不会产生任何错误，因为语句 y = 5 创建了一个全局命名空间。

代码：

```
x = 20
y = 5
x += y
print(x)
```

输出：

```
25
>>>
```

1.5.3 局部命名空间

查看以下示例代码。

代码：

```
x = 20
def print_x():
    x = 10
    print('Local variable x is equal to',x )
print('Global variable x is equal to',x )
print_x()
```

输出：

```
Global variable x is equal to 20
Local variable x is equal to 10
>>>
```

调用该函数时，Python 解释器尝试定位名为 x 的局部变量。如果没有，在全局命名空间中查找 x。

代码：

```
x = 20
def print_x():
    print('Local variable x is equal to',x )

print('Global variable x is equal to',x )
print_x()
```

输出：

```
Global variable x is equal to 20
Local variable x is equal to 20
>>>
```

局部变量，即变量在调用该函数时创建。无论何时调用函数，都会创建新作用域，并将变量分配给该作用域。一旦执行完函数，作用域就消失。在前面的示例代码中，调用函数 print_x()，在局部命名空间中找到局部变量 x = 10 并使用。该 x 值存在函数 print_x() 中，在该函数执行结束后随函数一起消失。

在函数命名空间中使用全局变量，需要使用该变量的 global 关键字来明确只使用全局变量。

```
x = 20
def print_x():
    global x
    x = 10
    print('Local variable x is equal to',x )

print('Global variable x is equal to',x )
print_x()
print('Global variable x is equal to',x )
```

一旦使用 global 关键字，函数 print_x() 就知道使用全局变量 x。在下条语句 x = 10 中，将 10 赋值给全局变量 x。因此，输出结果：调用函数前全局变量值为 20，调用函数 print_x() 后全局变量 x 的值为 10。函数使用带变量名的 global 关键字时，Python 不允许在同一函数中创建另一个同名变量，如图 1.2 所示。

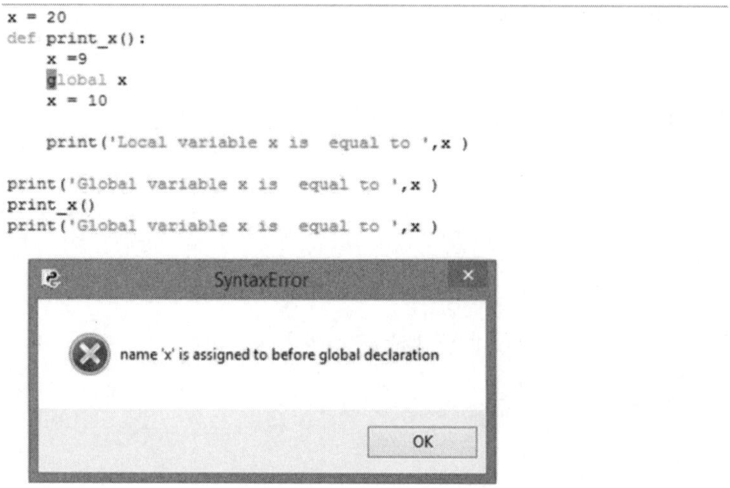

图 1.2

1.6 lambda 函数

lambda 函数为 Python 匿名函数，即定义时没有名称。这类函数用 lambda 关键字定义而非 def 定义。lambda 函数语法如下：

```
lambda arguments: expression
```

lambda 函数的最大特点是函数的参数个数不限，但只包含一个表达式。对该表达式计算并返回计算结果值。理想情况下，如果需要函数对象，则使用 lambda 函数。

注意：函数对象是指将 Python 函数作为参数传递给其他函数。函数也可赋值给变量或作为元素存储在数据结构中。例如：

代码：

```
your_age = lambda yr_of_birth: 2021 - yr_of_birth
print(your_age(1956))
```

输出：

```
65
```

map()、filter()和 reduce()函数使 Python 函数编程方法更便利。Python 3 中已经停止使用函数 reduce()。这些函数可被列表推导式（见《Python 核心编程：从入门到实践（学与练）》第 5 章）或循环替代。

示例 1.4

通过函数 filter()使用 lambda 函数。

答：

```
number = [1,2,3,4,5,6,13,7,8,9,0]
odd_number = list(filter(lambda x:(x%2!=0),number))
print(odd_number)
```

同理，map()函数把同一函数作用于序列的每个元素，并返回修改后的列表。计算列表 [1, 2, 3, 4, 5] 中每个数的平方值，通常使用如下方法。

```
list1 = [1,2,3,4,5]
for element in list1:
    print(element**2)
```

使用 map()和 lambda 函数能获得相同结果：

```
print(list(map(lambda x: x**2,list1)))
```

示例 1.5

使用 lambda 函数重写以下代码。

```
def comparisonFunc(a,b):
    if a>b:
        return a
    else:
        return b
print(comparisonFunc(10,3))
```

答：

```
comparisonFunc = lambda a,b: a if a > b else b
print(comparisonFunc(10,3))
```

示例 1.6

使用 lambda 函数重写以下代码。

```
def addFunc(a,b):
    return a + b
print(addFunc(5,2))
```

答:

```
addFunc = lambda a,b: a + b
print(addFunc(5,2))
```

示例 1.7
给定下面的列表:

```
countries = ['India','Mauritius','France','Turkey','Kenya','Hungary']
```

使用 lambda 函数输出列表中每个字符串的长度。
答:

```
countries = ['India','Mauritius','France','Turkey','Kenya','Hungary']
print(list(map(lambda x: len(x),countries)))
```

示例 1.8
在 lambda 函数中使用 sort( ) 函数,并说明 sort( ) 和 sorted( ) 函数的区别。
答:
sort( ) 函数的语法格式如下。

```
list.sort(key = None, reverse = False)
```

sort( ) 函数通过比较项对列表元素排序。lambda 函数使参数 key 变得更通用。
给定国家名称列表:

```
countries = ['India','Mauritius','France','Turkey','Kenya','Hungary']
```

如果按字母顺序对其名称排序,可使用以下代码:

```
countries = ['India','Mauritius','France','Turkey','Kenya','Hungary']
countries.sort()
print(countries)
```

或者使用 sort( ) 函数:

```
countries.sort(key = lambda x:x[0] )
print(countries)
```

因此,重点是通过 lambda 函数得到每个元素 (x) 并按照字符串的第 1 个元素排序。
给定姓名列表,根据姓氏对姓名排序。
编写以下代码。

```
>>> names = ['Mahatma Gandhi','Jawaharlal Nehru','Subhash Chandra
bose','Rani Laxmi Bai','Chandra Shekhar Azaad','Sarojini Naidu']
>>> names.sort(key = lambda x:x.split()[-1])
>>> print(names)

['Chandra Shekhar Azaad', 'Rani Laxmi Bai', 'Mahatma Gandhi', 'Sarojini Naidu',
'Jawaharlal Nehru', 'Subhash Chandra bose']
```

代码通过 lambda x:x.split()[-1]来实现。其中，x.split()将每个元素分成单独词汇，即将包含两部分的姓名分成两个单词的列表，将包含3部分的姓名分成3个单词的列表。-1 是列表最后一个元素的索引。接着，通过代码 lambda x:x.split()[-1]处理列表中每个元素的最后一个单词。（注意：'Subhash Chandra bose'放在末尾是因为这个姓氏不是以大写 B 开头。先将姓氏开头改为大写 B，然后查看输出）

```
>>> names = ['Mahatma Gandhi','Jawaharlal Nehru','Subhash Chandra Bose',
'Rani Laxmi Bai','Chandra Shekhar Azaad','Sarojini Naidu']
>>> names.sort(key = lambda x:x.split()[-1])
>>> print(names)

['Chandra Shekhar Azaad', 'Rani Laxmi Bai', 'Subhash Chandra Bose', 'Mahatma
Gandhi', 'Sarojini Naidu', 'Jawaharlal Nehru']
```

使用 sorted()函数也能得到相同的结果。示例代码如下所示。

```
>>> names = ['Mahatma Gandhi','Jawaharlal Nehru','Subhash Chandra Bose',
'Rani Laxmi Bai','Chandra Shekhar Azaad','Sarojini Naidu']

>>> print(list(sorted(names, key = lambda x:x.split()[-1])))

['Chandra Shekhar Azaad', 'Rani Laxmi Bai', 'Subhash Chandra Bose', 'Mahatma
Gandhi', 'Sarojini Naidu', 'Jawaharlal Nehru']
```

sort()和 sorted()函数的区别在于 sort()函数只修改原列表，而 sorted()函数创建新列表，并包含给定列表的排序版。

示例1.9

给定 List 1 = [(1, 2), (4, 1), (9, 10), (13, -3)]，编写代码获得以下输出。
(1) [(1, 2), (4, 1), (9, 10), (13, -3)]
(2) [(13, -3), (4, 1), (1, 2), (9, 10)]
(3) [(1, 2), (4, 1), (9, 10), (13, -3)]
(4) [(9, 10), (1, 2), (4, 1), (13, -3)]

答：
下面的代码对由元组构成的列表排序。x[0]表示根据位置 0 对元素排序；x[1]

表示根据位置 1 对元素排序。默认情况下排序按升序进行。如果按降序排序，则应将 reverse 设置为 True。

(1) 代码：

```
list1 = [(1,2),(4,1),(9,10),(13,-3)]
list1.sort(key = lambda x:x[0])
print(list1)
```

或

```
list1 = [(1,2),(4,1),(9,10),(13,-3)]
print(list(sorted(list1, key = lambda x:x[0])))
```

输出：

```
[(1,2),(4,1),(9,10),(13,-3)]
```

(2) 代码：

```
list1 = [(1,2),(4,1),(9,10),(13,-3)]
list1.sort(key = lambda x:x[1])
print(list1)
```

或

```
list1 = [(1,2),(4,1),(9,10),(13,-3)]
print(list(sorted(list1, key = lambda x:x[1])))
```

输出：

```
[(13,-3),(4,1),(1,2),(9,10)]
```

(3) 代码：

```
list1 = [(1,2),(4,1),(9,10),(13,-3)]
list1.sort(key = lambda x:x[0], reverse = True)
print(list1)
```

或

```
list1 = [(1,2),(4,1),(9,10),(13,-3)]
print(list(sorted(list1, key = lambda x:x[0], reverse = True)))
```

输出：

```
[(13,-3),(9,10),(4,1),(1,2)]
```

(4) 代码：

```
list1 = [(1,2),(4,1),(9,10),(13,-3)]
list1.sort(key = lambda x:x[1], reverse = True)
print(list1)
```

或

```
list1 = [(1,2),(4,1),(9,10),(13,-3)]
print(list(sorted(list1, key = lambda x:x[1], reverse = True)))
```

输出：

```
[(9,10),(1,2),(4,1),(13,-3)]
```

示例 1.10

函数 func() 定义如下。

```
def func(x,y):
    return (x+y)/2
```

使用 lambda 表达式重写函数。

答：

```
func1 = lambda x,y:(x+y)/2
```

示例 1.11

Alex 的成绩如下。

```
marks = [{'Subject':'Maths','Score':90},{'Subject':'Science','Score': 100},
{'Subject':'Geography','Score':83}]
```

使用 lambda 表达式按分数对成绩降序排序。

答：

代码：

```
marks = [{'Subject':'Maths','Score':90},{'Subject':'Science','Score': 100},
{'Subject':'Geography','Score':83}]
print(list(sorted(marks, key = lambda x:x['Score'], reverse = True)))
```

输出：

```
[{'Subject':'Science','Score': 100},{'Subject':'Maths','Score': 90},{'Subject':'Geography','Score': 83}]
```

示例 1.12

给定下面的列表：

```
list1 = ['AR-MO-UR','O-F','G-O-D']
```

使用 map() 函数和 lambda 表达式生成以下输出：

['ARMOUR','OF','GOD']

答：

```
>>> list1 = ['AR-MO-UR','O-F','G-O-D']
>>>print(list(map(lambda x:''.join(x.split('-')),list1)))
['ARMOUR','OF','GOD']
```

示例 1.13

list1 = [10, 30, 50, 70]，list2 = [20, 40, 60, 80]，编写代码输出 list1[i] + list2[i] 的和。其中，i 是元素索引。

答：

```
>>> list1 = [10,30,50,70]
>>> list2 = [20,40,60,80]
>>>print(list(map(lambda x,y: x+y,list1,list2)))
[30,70,110,150]
```

1.7 递归

编程就是解决复杂问题，好的程序员不但知道解决问题的方法不止一种，还应该掌握解决问题的多种技术。本节介绍递归，它是计算机编程基本技术，用于解决特定类型的复杂问题。

递归可从与原问题相似的、规模较小的问题推断出解决方案，也就是说，递归是指函数先解决与原问题相似的规模较小的问题，根据这个结果再计算得到最终结果。学习本节示例可以更加清晰地理解递归的定义。

1.7.1 递归求数的阶乘

数的阶乘是指所有小于及等于该数的正整数的积。5 的阶乘，即 5! 表示如下：

$$5! = 5 \times 4 \times 3 \times 2 \times 1 \tag{1}$$

仔细阅读前面的定义。因为数的阶乘是所有小于及等于该数的正整数的积，存在 $4 \times 3 \times 2 \times 1 = 4!$，所以式 (1) 可以重写为

$$5! = 5 \times 4 \times 3 \times 2 \times 1 = 5 \times 4! \tag{2}$$

根据阶乘的定义，式 (2) 可以写重写为

$$5! = 5 \times 4 \times 3 \times 2 \times 1 = 5 \times 4 \times 3! \tag{3}$$

同理，式 (3) 可以重写为

$$5! = 5 \times 4 \times 3 \times 2 \times 1 = 5 \times 4 \times 3 \times 2! \tag{4}$$

因此，如果定义 find_factorial(n) 函数计算数 n 的阶乘，意味着 find_factorial(n) 与 n * find_factorial(n-1) 相同，find_factorial(n-1) 与 (n-1) * find_factorial(n-2) 相同，依此类推，如图 1.3 所示。

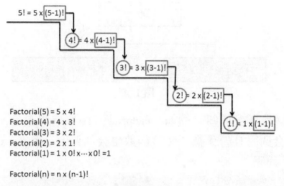

图 1.3

下面为递归代码的正确编写方法，但并不完整。

```
def find_factorial1(n):
    return n * find_factorial(n-1)
```

在递归中，找到终止条件很重要。递归的每个问题都有终止条件。它是不能在函数中继续使用递归的一个或多个特殊值，是递归函数的底层，也是最小计算部分。如果没有终止条件，则递归函数不会停止执行。递归函数逐步对自身进行调用，最后函数停止执行。

前面的代码因为没有终止条件，所以不完整。

根据阶乘定义，数的阶乘是所有小于及等于该数的正整数的积。这表示最后要乘的数永远为 1，因此，调用值 n=1 时函数停止执行。这就是 find_factorial() 函数的终止条件。完整代码如下所示。

```
def find_factorial(n):
    if(n==1):
        return 1
    return n * find_factorial(n-1)

print(find_factorial(5))
```

1.7.2 递归求阶乘的算法

函数 find_factorial（n）的执行步骤如下。

步骤 1：读取要求其阶乘的数 n。

步骤 2：判断读取的值是否等于 1。如果为真，则返回 1。

步骤 3：否则返回 n * find_factorial（n-1）的值。

1.7.3 递归的类型

递归分为直接递归和间接递归，如图1.4所示。

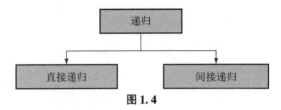

图1.4

函数调用本身就是直接递归。例如，factorial()函数调用本身，因此称为直接递归。可是，f()函数有时会调用f1()函数，f1()函数也会回调f()函数。这是间接递归，如图1.5所示。参看以下代码。

```
def happy_new_year(n=1):
    if(n<=0):
        how_many_times()

    for i in range(n):
        print("Happy New Year")

def how_many_times():
    val = input("How many times should I print?:")
    if val =='':
        happy_new_year()
    else:
        happy_new_year(int(val))

how_many_times()
```

在前面的示例代码中，happy_new_year()函数输出Happy New Year，次数不限。如果不提供值，即默认值为1，则输出1次消息。如果值无效，即0或更小，则调用how_many_times()函数，该函数提示用户再次输入新值，并使用新值回调happy_new_year()函数。

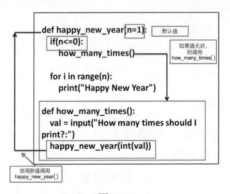

图1.5

以下是这段代码的几个有趣的输出。

实例1：调用 happy_new_year( ) 函数时无输入值。

输出：

```
Happy New Year
```

说明：没有提供值，根据默认值消息只输出 1 次。

实例2：当 n = 0 时调用 happy_new_year( ) 函数。

输出：由于 n = 0，再次调用 how_many_times( ) 函数，直到提供一个大于 0 的 n 值。

```
How many times should I print?:0
How many times should I print?:0
How many times should I print?:7
Happy New Year
Happy New Year
Happy New Year
Happy New Year
Happy New Year
Happy New Year
Happy New Year
>>>
```

说明：用户输入两次 n = 0，因此 happy_new_year( ) 函数调用 how_many_times( ) 函数，直至用户输入一个大于 0 的值，示例中为 7。因此，当 n <= 0 时，函数提示用户重新输入值，并使用新输入值再次调用 happy_new_year( ) 函数。

实例3：多次使用无效值调用。

输出：

```
happy_new_year(-2)
How many times should Iprint?:-3
How many times should I print?:-8
How many times should I print?:0
How many times should I print?:2
Happy New Year
Happy New Year
```

说明：将无效值传递给 happy_new_year( ) 函数，接着调用 how_many_times( ) 函数，提示用户输入有效值。用户仍提供无效值，两个函数就继续相互调用，直到输入有效值。

1.7.4 递归的优、缺点

下面分析递归的优、缺点。

1）递归的优点

（1）代码行更少，代码更简洁。

（2）可将复杂工作拆分成简单工作。

2）递归的缺点

（1）不易培养递归逻辑思维。

（2）调试递归函数较为困难。

（3）递归函数会消耗更多内存和时间。

示例 1.14 （斐波那契数列的递归）

斐波那契数列的整数排列如下：

0，1，1，2，3，5，8，13，21，34，55，89，144，…

编写代码前，理解序列的创建规律非常重要，如图 1.6 所示。

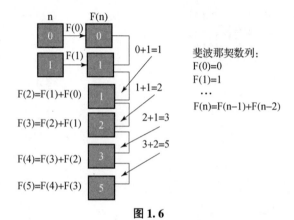

图 1.6

根据图 1.6 中的逻辑，很容易计算出：

$$F(6) = F(5) + F(4) = 5 + 3 = 8$$
$$F(7) = F(6) + F(5) = 8 + 5 = 13$$

因此，$F(n) = F(n-1) + F(n-2)$。

通过以下步骤创建斐波那契数列。

Function fib(n)：

步骤1：读取数值 n。

步骤2：如果 n=0，返回 0。

步骤3：如果 n=1，返回 1。

步骤4：否则返回 fib(n-1) + fib(n-2)。

代码如下所示。

```
def fib(n):
    if(n==0):
        return 0
    elif(n==1):
        return 1
    else:
        return fib(n-1)+fib(n-2)
```

1.8 存储

通常情况下，利用存储技术维护查找表时，如果其中存储了解决方法，就不必反复解决同一个子问题；否则，解决并存储这些值就可重复利用。

例如，斐波那契数列如下：

```
F(n) = F(n-1) + F(n-2)(n>1)
     = n(n = 0,1)
```

因此，

```
F(n):
    if n < 1:
        return n
    else:
        return F(n-1) + F(n-2)
```

这里进行两个递归调用，将相加后返回该值，如图 1.7 所示。

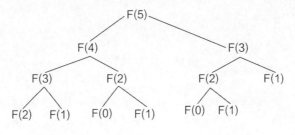

图 1.7

注意，只为找到 fibonacci(5)，fibonacci(2) 就计算了 3 次，fibonacci(3) 计算了 2 次。因此，随着 n 的增大，fibonacci() 函数的 s(f(n)) 性能也会下降。时间和空间的消耗也会随着 n 的增大呈指数增长。但能通过一种简便方法节省时间，即在第 1 次计算时存储该值，也就是第 1 次计算 F(1)、F(2)、F(3) 和 F(4) 时存储它们的值，依此类推。编写以下代码。

```
F(n):
    if n <= 1:
        return n
    elif F(n) exist :
        return F(n-1)
    else:
        F(n) = F(n-1) + F(n-2)
        Save F(n)
        Return F(n)
```

对以上代码中说明如下。

（1）由于斐波那契数列从 0 开始，fibonacci()函数接收值并创建一个大小为 num + 1 的列表 fib_num。

（2）调用 fib_calculate()函数接收值 num，并将列表 fib_num 作为参数。

（3）列表中所有索引都存储 -1。

①如果 fib_num[num] > 0，表示该数的斐波那契值已存在，则不用再次计算，直接返回该数。

②如果 num <= 1，则返回 num。

③如果 num >= 2，则计算 fib_calculate(num - 1, fib_num) + fib_calculate(num - 2, fib_num)。计算结果需要存储在列表 fib_num 的 num 索引处，以便不用时再次计算。

代码：

```python
def fibonacci(num):
    fib_num = [-1]*(num + 1)
    return fib_calculate(num, fib_num)
def fib_calculate(num, fib_num):
    if fib_num[num] >= 0:
        return fib_num[num]
    if (num <= 1):
        fnum = num
        return fnum
    else:
        fnum = fib_calculate(num-1, fib_num) + fib_calculate(num-2, fib_num)
        fib_num[num] = fnum
        return fnum

num = int(input('Enter the number: '))
print("Answer = ",fibonacci(num))
```

执行：

```python
num = int(input('Enter the number: '))
print("Answer = ",fibonacci(num))
```

输出：

```
Enter the number: 15
Answer = 610
>>>
```

知识要点

- 函数是能多次调用的封装代码块。
- 函数可简化编码且优点众多。
- 每个函数均用于执行单个相关操作。
- Python编程中函数分为：
  - Python提供的内置函数；
  - 为执行特定任务而开发、非Python提供的用户自定义函数。
- 函数块以def关键字开头，后面为函数名和括号"()"，括号"()"后面跟冒号":"，以标记函数代码块开头。
- 代码块需要按照PEP-8正确缩进。
- 函数名后跟括号可随时调用函数。
- 函数不会执行显式调用。
- docstring（文档字符串）可记录函数的重要特性。
- 函数能否接收参数取决于函数定义。
- 参数是函数的局部变量，是定义的一部分。
- Python中有5种函数参数。
  - 位置参数；
  - 默认参数；
  - 关键字参数；
  - *args；
  - **kwargs。
- return语句将表达式返回给调用者并退出函数。
- 程序可能不返回任何内容（None）。
- 命名空间是包含程序中所有定义命名（变量/函数/类）的空间。
- 命名空间可确保名称的唯一性，以防存在冲突。
- Python中的命名空间通过字典实现。
- 命名空间维护命名到对象的映射，其中命名是键，对象是值。
- 命名或变量存在于代码的特定区域中，该区域定义了其作用域。变量可到达其作用域中的任意位置。命名空间或作用域的三种类型如下。
  - 内置命名空间：其中的内置函数可跨越所有文件或模块使用。
  - 全局命名空间：包含在单个文件中可用的所有变量、函数和类。
  - 局部命名空间：函数定义的变量。
- 递归依据本身定义。递归是指函数反复调用自身，直到满足终止条件才停止执行。
- 函数调用自身称为递归。相同指令集重复执行，确定递归调用终止条件非常重要。

- 递归分为两种类型：直接递归和间接递归。
- 递归函数使代码更简洁。
- 递归消耗更多内存和时间，并且调试困难。
- 递归函数的两个要素：终止条件、递归条件。
- 递归函数使用迭代来实现，反之亦然。

小结

本章介绍了创建函数的方法。现在读者已经具备使用面向对象编程所需的所有知识。下一章将介绍创建 Python 类的方法。

简答题

1. 可重用并按其命名调用的代码块称为_____。

答：函数

2. 为什么需要函数？

答：通常程序中的某些指令可以多次调用。与其需要时反复编写同一代码，不如定义函数，将代码放入其中在需要时调用此函数即可。这样不仅节省时间和精力，还可简化开发程序。函数既有助于组织编码工作，也让代码测试变得很容易。

3. 函数传递参数中不可变对象和可变对象有哪些区别？

答：将字符串、整数或元组等不变参数传递给函数，则引用传递对象，其参数值无法更改，就像按值传递调用。可变对象按对象引用重新传递，其值可以更改。

4. 导入代码中定义内置函数的模块，可以轻松使用内置函数。

a. 正确

b. 错误

答：b

5. 函数能提供更好的模块化。

a. 正确

b. 错误

答：a

编程题

1. 编写代码查找字符串中的所有回文。

答：

查找字符串中所有回文的步骤如下。
（1）创建所有可能的子字符串的列表。
（2）使用 for 循环将字符串切分为所有可能的级别，从而创建子字符串。
（3）检查每个子字符串是否为回文。
（4）将子字符串转换为单个字符的列表。
（5）按相反顺序，将列表中的字符添加到字符串中。
（6）如果结果字符串与原始字符串匹配，则它是回文。

代码：

```
def create_substrings(x):
    substrings = []
    for i in range(len(x)):
        for j in range(1, len(x)+1):
            if x[i:j] !='':
                substrings.append(x[i:j])
    for i in substrings:
        check_palin(i)
def check_palin(x):
    palin_str = ''
    palin_list = list(x)
    y = len(x)-1
    while y>=0:
        palin_str = palin_str + palin_list[y]
        y = y-1
    if(palin_str == x):
        print("String ",x," is a palindrome")
x = "malayalam"
create_substrings(x)
```

执行：

```
x = "malayalam"
create_substrings(x)
```

输出：

```
String  m  is a palindrome
String  malayalam  is a palindrome
String  a  is a palindrome
String  ala  is a palindrome
String  alayala  is a palindrome
```

```
String  l    is a palindrome
String  layal is a palindrome
String  a    is a palindrome
String  aya  is a palindrome
String  y    is a palindrome
String  a    is a palindrome
String  ala  is a palindrome
String  l    is a palindrome
String  a    is a palindrome
String  m    is a palindrome
```

2. 下面函数的输出结果是什么?

```
def happyBirthday():
    print("Happy Birthday")
a = happyBirthday()
print(a)
```

答:

```
Happy Birthday
None
```

3. 下面代码的输出结果是什么?

```
def outerWishes():
    global wishes
    wishes = "Happy New Year"
def innerWishes():
    global wishes
    wishes = "Have a great year ahead"
    print('wishes =', wishes)
wishes = "Happiness and Prosperity Always"
outerWishes()
print('wishes =', wishes)
```

答:

```
wishes = Happy New Year
```

4. 下面代码的输出结果是什么?

```
total = 0
def add(a,b):
    total = a+b
```

```
        print("inside total = ",total)
add(6,7)
print("outside total = ",total)
```

答:

```
inside total = 13
outside total = 0
```

5. 编写使用欧几里得算法查找 HCF 的函数，并返回该值。

答:

图 1.8 所示为查找 HCF 的两种方法。其中，左侧是查找 HCF 的传统方法，右侧是使用欧几里得算法查找 HCF 的方法。

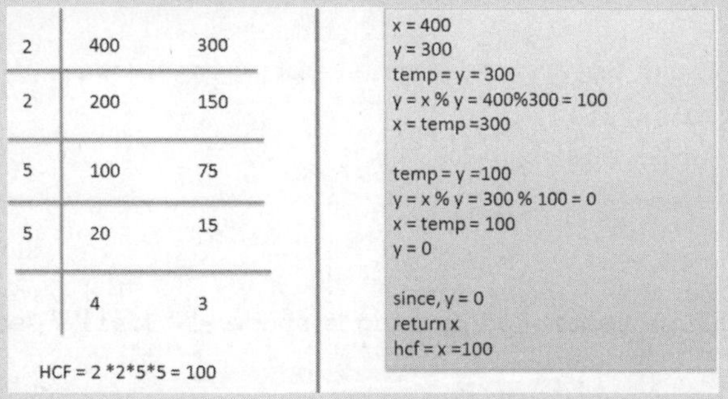

2	400	300
2	200	150
5	100	75
5	20	15
	4	3

HCF = 2 *2*5*5 = 100

```
x = 400
y = 300
temp = y = 300
y = x % y = 400%300 = 100
x = temp =300

temp = y =100
y = x % y = 300 % 100 = 0
x = temp = 100
y = 0

since, y = 0
return x
hcf = x =100
```

图 1.8

代码:

```
def hcf(x,y):
    small_num = 0
    greater_num = 0
    temp = 0
    if x > y:
        small_num = y
        greater_num = x
    else:
        small_num = x
        greater_num = y
    while small_num > 0:
        temp = small_num
        small_num = greater_num % small_num
        greater_num = temp
    return temp
```

执行：

```
print("HCF of 6 and 24 = ",hcf(6,24))
print("HCF of 400 and 300 = ",hcf(400,300))
```

输出：

```
HCF of 6 and 24  = 6
HCF of 400 and 300  = 100
```

6. 证明不返回函数会隐式返回。

答：

```
def find_len(list1):
    print("List Received")

x = input('Enter the values separated by single space :')
x_list = x.split()
print(find_len(x_list))
```

输出：

```
Enter the values separated by single space :1 'Gmail''Google'1.09
[2,3,45,9]
5
List Received
None
>>>
```

说明： find_len()函数不返回列表长度；print 语句不显示内容。因此，证明了函数不显式返回时隐式返回 None。

7. 写出下列代码的输出结果。

```
x = 20
def print_x():
    global x
    x = 10
    return x
print(print_x())
```

答： 10。

8. 写出下列代码的输出结果。

```
x = 20
def function_1():
    global x
    x = 10
    return x
def function_2():
    return x

print(function_2())
```

答：20。

9. 编写函数，使用 for 循环计算数的阶乘。

答：

使用 for 循环计算数的阶乘代码如下。

代码：

```
def factorial(number):
    j = 1
    if number ==0 | number ==1:
        print(j)
    else:
        for i in range (1, number +1):
            print(j," * ",i," = ",j * i)
            j = j * i
    print(j)
factorial(5)
```

执行：

```
factorial(5)
```

输出：

```
1 * 1 = 1
1 * 2 = 2
2 * 3 = 6
6 * 4 = 24
24 * 5 = 120
120
```

10. 使用 for 循环编写求斐波那契数列的函数。

答：

斐波那契数列：0，1，1，2，3，5，8，…。

取三个变量：i，j 和 k。

(1) if i = 0，j =0，k =0

(2) if i =1，j =1，k =0

(3) if i > 1：

    temp = j

    j = j + k

    k = temp

计算结果如表 1.1 所示。

表 1.1

i	k	j
0	0	0
1	0	1
2	0	temp = j = 1 j = j + k = 1 + 0 = 1 k = temp = 1
3	1	temp = j = 1 j = j + k = 1 + 1 = 2 k = temp = 1
4	1	temp = j = 2 j = j + k = 2 + 1 = 3 k = temp = 2
5	2	temp = j = 3 j = j + k = 3 + 2 = 5 k = temp = 3
6	3	temp = j = 5 j = j + k = 5 + 3 = 8 k = temp = 1

代码：

```python
def fibonacci_seq(num):
    i = 0
    j = 0
    k = 0
    for i in range(num):
        if i ==0:
```

```
            print(j)
        elif i ==1:
            j = 1
            print(j)
        else:
            temp = j
            j = j + k
            k = temp
            print(j)
```

执行：

```
fibonacci_seq(10)
```

输出：

```
0
1
1
2
3
5
8
13
21
34
```

11. 使用 while 循环改写下面代码。

```
def test_function(i,j):
    if i == 0:
        return j
    else:
        return test_function(i -1,j +1)
print(test_function(6,7))
```

答：

```
def test_function(i,j):
    while i > 0:
        i = i - 1
        j = j +1
    return j
print(test_function(6,7))
```

12. 下面代码的输出结果是什么？

```
total = 0
def add(a,b):
    global total
    total = a+b
print("inside total = ",total)
add(6,7)
print("outside total = ",total)
```

答：
输出结果如下所示。

```
inside total = 13
outside total = 13
```

13. 编写代码，使用递归查找从0到给定数字的自然数之和。

答：
结果如表1.2所示。

表1.2

i	结果
0	0
1	1 + 0 = i(1) + i(0) = 1
2	2 + 1 = i + i(1) = 3
3	3 + 3 = i + i(2) = 6
4	4 + 6 = i + i(3) = 10
5	5 + 10 = i + i(4) = 15

观察 i=0，结果为0，然后结果 = i(n) + i(n-1)。

代码：

```
def natural_sum(num):
    if num == 0:
        return 0
    else:
        return (num + natural_sum(num-1))
```

执行：

```
print(natural_sum(10))
```

输出：

55

14. 下列代码的输出结果是什么？

```
def funny(x,y):
    if y == 1:
        return x[0]
    else:
        a = funny(x, y-1)
        if a > x[y-1]:
            return a
        else:
            return x[y-1]
x = [1,5,3,6,7]
y = 3
print(funny(x,y))
```

答：5。

说明：为了更好地理解下面代码中显示的打印语句，再次执行代码。代码执行顺序如下。

```
def funny(x,y):
    print("calling funny , y = ",y)
    if y == 1:
        return x[0]
    else:
        print("inside else loop because y = ", y)
        a = funny(x, y-1)
        print("a = ",a)
        if a > x[y-1]:
            print("a = ",a, " Therefore a > ",x[y-1])
            return a
        else:
            print("a = ",a, " Therefore a < ",x[y-1])
            return x[y-1]
x = [1,5,3,6,7]
y = 3
print(funny(x,y))
```

输出：

```
calling funny , y = 3
inside else loop because y = 3
calling funny , y = 2
inside else loop because y = 2
calling funny , y = 1
```

```
a = 1
a = 1  Therefore a < 5
a = 5
a = 5  Therefore a > 3
5
The answer is 5
```

15. 下面代码的输出结果是什么？

```
def funny(x):
    if (x% 2 == 1):
        return x +1
    else:
        return funny(x -1)
print(funny(7))
print(funny(6))
```

答：
对于 x =7：
(1) x = 7；
(2) x % 2 is 1；
(3) return 7 + 1 = 8。
对于 x = 6：
(1) x =6；
(2) x%2 = 0；
(3) return funny (5)。
对于 x = 5：
(1) x%2 =1；
(2) return 5 +1 = 6。

16. 使用递归方法编写代码，求斐波那契数列。

答：
斐波那契数列 = 0, 1, 2, 3, 5, 8, 13, …。
结果如表1.3所示。

表1.3

i	结果
0	0
1	1
2	1 +0 = i(0) + i(1) = 1
3	1 +1 = i(2) + i(1) = 2
4	2 +1 = i(3) + i(2) = 3
5	3 +2 = i(4) + i(3) = 5

观察 i = 0，结果为 0；i = 1，结果为 1。此后，i(n) = i(n-1) = i(n-2)。使用递归方法求斐波那契数列也能得到相同的结果。

（1）fibonacci_seq(num)将数字作为参数。

（2）如果 num = 0，结果为 0；

如果 num = 1，结果为 1；

否则结果是 fibonacci_seq(num-1) = fibonacci_seq(num-2)。

（3）如果计算 10 的斐波那契数列，那么

①元素为 0~10。

②调用 fibonacci_seq( ) 函数；

- fibonacci_seq(0) = 0；
- fibonacci_seq(1) = 1；
- fibonacci_seq(2) = fibonacci_seq(1) + fibonacci_seq(0)；
- fibonacci_seq(3) = fibonacci_seq(2) + fibonacci_seq(3)。

代码：

```
def fibonacci_seq(num):
    if num < 0:
        print("Please provide a positive integer value")
    if num == 0:
        return 0
    elif num == 1:
        return 1
    else:
        return ( fibonacci_seq(num-1) + fibonacci_seq(num-2) )
```

执行：

```
for i in range(10):
    print(fibonacci_seq(i))
```

输出：

```
0
1
1
2
3
5
8
13
21
34
```

17. 下面代码的输出结果是什么？

```
def test_function(i,j):
    if i == 0:
        return j
    else:
        return test_function(i-1,j+1)
print(test_function(6,7))
```

答：13。

说明如表 1.4 所示。

表 1.4

i	j	i == 0？	返回值
6	7	No	test_function(5,8)
5	8	No	test_function(4,9)
4	9	No	test_function(3,10)
3	10	No	test_function(2,11)
2	11	No	test_function(1,12)
1	12	No	test_function(0,13)
0	13	Yes	13

18. 下面代码的输出结果是什么？

```
def even(k):
    if k <= 0:
        print("please enter a positive value")
    elif k == 1:
        return 0
    else:
        return even(k-1) + 2
print(even(6))
```

答：10。

说明如表 1.5 所示。

表 1.5

k	k <= 0	k == 1	结果
6	No	No	even(5) +2
5	No	No	even(4) +2 +2
4	No	No	even(3) +2 +2 +2
3	No	No	even(2) +2 +2 +2 +2
2	No	No	even(1) +2 +2 +2 +2 +2
1	No	Yes	0 +2 +2 +2 +2 +2 = 10

19. 编写代码，用递归方法求 3 的 n 次方。

答：

（1）定义函数 n_power(n)，将乘方的值作为参数（n）。

（2）如果 n = 0，则返回 1，因为任何幂为 0 的数都是 1；否则返回(n_power(n − 1))。

结果如表 1.6 所示。

表 1.6

n	n < 0	n == 0	结果
4	No	No	n_power(3) *3
3	No	No	n_power(2) *3 *3
2	No	No	n_power(1) *3 *3 *3
1	No	No	n_power(0) *3 *3 *3 *3
0	No	Yes	1 *3 *3 *3 *3

代码：

```
def n_power(n):
    if n < 0:
        print("please enter a positive value")
    elif n == 0:
        return 1
    else:
        return n_power(n - 1) *3
print(n_power(4))
```

执行：

```
print(n_power(4))
```

输出：
```
81
```

1. 简述函数。

答：

提示如下。

（1）函数定义。

（2）函数在 Python 程序设计中的重要性。

（3）Python 的函数分类或类型及示例。

2. 简述函数调用。

答：

调用函数的语法格式如下：

```
function_name(parameters)
```

其中，function_name 是调用的函数名称，参数是函数根据其定义所需的值，所有参数必须用逗号分隔，并且用括号括起来。在调用函数时，参数映射到相应参数，并执行相应代码块。

3. 形参和实参的区别。

答：

函数定义形参（也称为参数），放在括号内。调用函数时形参接收实际值。

实参是函数调用中存在的参数，将其值传递给函数。必须按照函数定义传递值。在调用函数时，形参被实参替换。

4. Python 包括哪些类型的函数？

答：

Python 有两种类型的函数。

（1）内置函数：Python 中的库函数。

（2）用户自定义函数：由程序开发人员定义。

5. 什么是函数头？

答：

函数定义的第 1 行以 def 开头，以冒号（:）结尾，称为函数头。

6. 函数何时执行？

答：

函数在调用时执行。可以直接从 Python 提示符或其他函数中调用。

7. 什么是形参？形参和实参的区别是什么？

答：

形参是在函数定义中定义的变量，而实参是传递给函数的实际值。实参包含的数据传递给形参。形参可以按名称传递。

```
def function_name(param):
```

在上面语句中，param 是形参。下面的语句演示了函数调用过程。

```
function_name(arg):
```

其中，arg 是调用函数时传递的数据。在此语句中，arg 是实参。

因此，形参只是方法定义中的变量，实参是调用函数时传递给方法形参的数据。

8. 打印值的函数和返回值的函数的区别是什么？

答：

函数返回值时会返回能捕获的值，换句话说，将返回值分配给变量，而打印该值的函数却不同。

9. return 语句的用途是什么？

答：

return 语句可以退出函数，并将值返回给函数的调用者。在下面代码中可以看到，函数 func() 返回两个数的和并将此值赋给 total，然后打印 total 的值。

```
def func(a,b):
    return a + b

total = func(5,9)
print(total)
```

10. 局部范围嵌套在全局范围内，全局范围嵌套在内置范围内。这句话是什么意思？

答：

变量作用域表示从代码的哪个部分使用程序。变量范围可以是局部，也可以是全局。

局部变量在函数内部定义，全局变量在函数外部定义。局部变量只能在定义它们的函数中访问。所有函数都可以在整个程序中访问全局变量。示例代码如下。

```
total = 0 # 全局变量
def add(a,b):
    sumtotal = a + b # 局部变量
    print("inside total = ",total)
```

在引用变量时，Python 首先在其范围内查找该变量，如果找不到该变量，将在所有封闭范围和命名空间中查找。

11. 什么是局部变量和全局变量？

答：

局部变量在函数中定义和使用。解释器不识别在该函数之外的函数中定义的任何变量，只有在执行定义该变量的函数时才能识别到该变量，并且在该函数结束时，该变量标识符被释放。全局变量如同其名，在文件的任何地方都可使用。在函数内部或外部使用全局变量时，无须再次声明。

变量 x 是全局变量，文件中的任何代码块均可使用，但变量 a、b 和 c 不同，只能在定义它们的代码块中使用，因此变量 a、b 和 c 是局部变量，如图 1.9 所示。

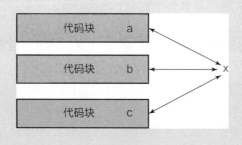

图 1.9

12. 什么是默认参数？

答：

默认参数也称为可选参数。在定义函数时，如果参数提供默认值，则称为默认参数。如果用户在调用函数时不为该参数提供任何值，则函数使用函数定义中的默认值。

13. 定义任意 3 种类型的 Python 函数参数。

答：

Python 函数参数有以下 3 种类型。

（1）默认参数：如果用户不提供值，则使用默认值。示例代码如下。

```
def func(name = "Angel")
print("Hppy Birthday",name)

func()
Happy Birthday Angel
```

name 的默认值是 Angle，因为用户不提供任何值，所以使用默认值。

（2）关键字参数：调用函数并传递值，不考虑其位置，调用函数时使用参数名并为其赋值。示例代码如下。

```
def func(name1, name2):
    print("Happy Birthday", name1, " and ",name2,"!!!")

func(name2 = "Richard",name1 = "Marlin")
```

输出：

```
Happy Birthday Marlin and Richard !!!
```

（3）可变长度参数：如果不确定函数的参数个数，则使用可变长度参数。函数定义中在参数前放置"＊"，这样从这点到末尾的所有位置参数都作为元组。另外，如果将"＊＊"放在参数名称前面，则表示从该点到末尾的所有位置参数都作为字典。示例代码如下。

```
def func( *name, **age):
    print(name)
    print(age)
func("Lucy","Aron","Alex", Lucy = "10",Aron = "15",Alex = "12")
```

输出：

```
('Lucy','Aron','Alex')
{'Lucy':'10','Aron':'15','Alex':'12'}
```

14. 什么是结果函数和非结果函数？

答：

结果函数是返回值的函数，而非结果函数是不返回值的函数。非结果函数也称为空函数。

15. 什么是匿名函数？

答：Python 的 lambda 工具用于创建无名称函数。这类函数也称为匿名函数。lambda 函数是非常小的函数，函数体中只有一行语句，并且没有返回语句。示例代码如下。

```
total = lambda a, b: a + b
total(10,50)
60
```

第 2 章

类、对象与继承

> **引言** 本章介绍面向对象编程,也称为 OOP。到目前为止,读者要做的就是使用变量、函数与流控制创建一组预定义指令。本章展示 Python 也是可以使用类、对象、属性和方法的面向对象的编程语言。实际上,Python 是多范式编程语言,支持命令式、过程式、函数式和面向对象的范式。因此,其充满灵活性,开发人员可根据具体情况选择想要使用的方法。

> **知识结构**
> - 类和对象
> - 析构函数__del__( )
> - 类变量的类型
> - 继承

> **目标** 完成本章的学习后,读者应掌握以下技能。
> (1) 创建类。
> (2) 为高级编程实现继承。
> 面向对象编程就是遵循建立与现实世界存在数据关联的对象的方法,然后对象之间以其现实世界中的关联方式互相关联。因此,学校管理系统程序中有班级、学生、教师、教学大纲、学科等对象,然后根据现实世界中它们之间的关系彼此联系。面向对象编程可使开发人员创建个人的现实世界数据类型。

2.1 类和对象

面向对象编程允许按照个人思维方式或规则设计代码。在处理小问题时,面向对象编程对问题的处理方式影响不大,但在实际软件开发行业中,项目通常非常复杂。如果不遵循正确的方法，在代码中包含每种可能性和关联可能非常困难。如果将实际场景分解成更小、更简单的部分，每次只专注某个部分，然后将这些部分相互关联以获得最终结果，这将更容易解决问题，并且将得到一个包含所有特性的代码。这些较小部分称为对象。编码概念与人类的思维方式非常相似。作为项目设计师，需要关注实体类的功能、工作原理及可用之处。每个对象都有状态和行为。为了正确理解对象概念，首先理解类和实例的概念。

  类只是创建对象所需的定义或蓝图。如果为一家公司搭建管理系统，则员工信息是这个系统中最重要的部分。因此，如果以 Employee 为名称创建一个类，则这个类将作为创建 Employee 对象的蓝图。此外，虽然实例和对象在现实世界中常互换使用，但理解两者之间的区别同样重要。对象表示类的内存地址，而实例只是类在特定情况下的虚拟副本。假设存在一个类 Employee，并有 6 条员工信息。每位员工均为不同对象，并且都有不同的个人信息，如姓名、电子邮件 id、年龄等。每个实例都是单一对象的唯一实际表现。对象使用实例执行操作。对象是通用的，但实例是特定的。每个实例都被赋值，同时使实例之间得以区分。因此，对象是通用的，但实例是类的实际表现。实例表示特定的对象，也表示在内存中创建的单个对象。这将在下一节中详细解释。

  Python 中的一切均为对象，都是类的实例。类是一种用户定义的数据类型。具有该数据类型的对象，称为该类的实例。

```
>>> x = 1
>>> type(x)
<class 'int'>
>>> y = "String"
>>> type(y)
<class 'str'>
```

  那么，类和对象之间有什么区别呢？对象是一组实例，其中作为实例引用一个特定的表示，如员工表（表 2.1）。

表 2.1

name	email	department	age	salary
Alex	alex@ company_name. com	Finance	42	500 000
Evan	evan@ company_name. com	HR	34	300 000
Maria	maria@ company_name. com	HR	30	350 000
Pradeep	pradeep@ company_name. com	IT	28	700 000
Simon	simon@ company_name. com	Finance	40	500 000
Venkatesh	venkatesh@ company_name. com	Sles	25	400 000

显然，员工具有更多详细信息，如电话号码、地址、性别、婚姻状况、银行账户等，但这里选取简单示例，以便抓住主题。现在，如果试图从表 2.1 中创建 Employee 类，则使用这个类创建的所有员工（Alex, Evan, Maria, Pradeep, Simon, Venkatesh）都是对象。然而，当讨论给定 Employee 类时，该 Employee 类的特定表示即实例。因此，表 2.1 中的每行均为一个实例。

现在，注意表 2.1 中的每一列：姓名、电子邮件 id、部门、年龄、薪水。这些列保存与员工相关的信息。保存与对象相关的数据信息的变量称为属性。

现在，创建具有以下属性的 Employee 类。

（1）name；
（2）email；
（3）department；
（4）age；
（5）salary。

Python 使用 class 关键字创建类，后跟类名。

创建类的语法格式如下：

```
class class_name:
```

class 关键字和其他模块定义一样，任何包含于类中的内容均为其一部分。在此条件下，讨论内容为 Employees。因此，可用如下代码定义 Employee 类。

```
class Employee:
```

注意：按照惯例，类名采用驼峰命名法，以便与 Python 提供的标准库类区分。驼峰命名法即类名中每个单词都以大写字母开头，如 **Employee**、**MyEmployee** 等。

与函数一样，类必须至少包含一条指令。即使创建空类，类中也应至少有一条指令。因此，空 Employee 类应如下所示。

```
class Employee:
    pass
```

其中，pass 是一个空指令，不执行，但可用于避免语法错误。

在定义类之后，可定义类的属性和方法。方法为类中定义的函数，它与给定的对象一起执行。方法需要访问使其执行的对象，因此，类中每个方法的第一个参数是当前实例，按照惯例，该实例被称为 self。在方法内部使用 self 来访问存储在对象中的数据。

Python 定义了一些名称以双下划线开始和结束的特殊方法（也称为函数）。最常用的是 __init__() 函数。此方法用于初始化类实例，也可称为类构造函数。因为每当 Python 创建该类实例时，会自动调用 __init()__ 函数，所以只需将其定义，而不必显式调用。除此之外，__init__() 函数并非类的强制性设置，而是可选的。下面定义 __init__() 函数。

```python
class Employee:
    def __init__(self, name, email, department, age, salary):
        self.name = name
        self.email = email
        self.department = department
        self.age = age
        self.salary = salary
```

下面逐一分析。

定义部分如下。

```
def __init__(self, name, email, department, age, salary):
```

(1) __()init__是一个特殊函数，为类的构造函数。
(2) 该函数的第一个参数为 self。
(3) self 将该方法连接到类的实例（调用该方法的对象）。
(4) self 后面的参数为方法的参数值。
(5) 一旦方法能够访问对象，方法参数的值将被传递给对象属性。

下面观察函数部分。

```
self.name = name
self.email = email
self.department = department
self.age = age
self.salary = salary
```

从上面的代码可见，"="号左边为对象属性，而右边为方法参数。按照惯例，对象属性和方法的参数使用相同名称。虽然方法的参数名称可以不同，但会使代码改变。例如：

```python
class Employee:
    def __init__(self, a, b, c, d, e):
        self.name = a
        self.email = b
        self.department = c
        self.age = d
        self.salary = e
```

在本书所举示例中，不存在打破惯例的情况。

接下来，学习__init__()函数的工作原理。在表 2.1 中，第一位员工的姓名为 Alex；其电子邮件 id 为 alex@ company_name. com；在财务部工作；42 岁；薪水为 500 000 元。

因此创建如下类实例：

```
e1 = Employee("Alex","alex@company_name.com","Finance",42,500000)
```

该语句调用__init__()函数。参数如下。
(1) self，将该方法连接到类的实例 e1。

(2)"Alex"。
(3)"alex@ company_name. com"。
(4)"Finance"。
(5)42。
(6)500 000。

现在，__init__()函数中作为参数传递的值被赋给对象属性。代码如下。

```
self.name = "Alex"
self.email = "alex@company_name.com"
self.department = "Finance"
self.age = 42
self.salary = 500000
```

通过这个单独实体，就创建了一个实例。

到目前为止，该类的代码如下。

```
class Employee:
    def __init__(self, name, email, department, age, salary):
        self.name = name
        self.email = email
        self.department = department
        self.age = age
        self.salary = salary

e1 = Employee("Alex","alex@company_name.com","Finance",42,500000)
```

如果执行这段代码，不会得到任何输出，因为已创建对象，但未被使用。

按如下方法输出对象值。

```
def print_employee(self):
    print("Name : ",self.name)
    print("Email : ",self.email)
    print("Department : ",self.department)
    print("Age : ",self.age)
    print("Salary : ",self.salary)
```

print_employee()方法只有 self 作为参数，这使之能访问对象（本示例中为 e1）。
可以调用与对象结合的方法，代码如下。

```
object_name.method_name()
```

在本示例中，对象名为 e1，方法名为 print_employee()。因此，可使用以下代码。

```
e1.print_employee()
```

现在代码如下。

```
class Employee:
    def __init__(self, name, email, department, age, salary):
```

```
        self.name = name
        self.email = email
        self.department = department
        self.age = age
        self.salary = salary

    def print_employee(self):
        print("Name : ",self.name)
        print("Email : ",self.email)
        print("Department : ",self.department)
        print("Age : ",self.age)
        print("Salary : ",self.salary)

e1 = Employee("Alex","alex@company_name.com","Finance",42,500000)
e1.print_employee()
```

注意最后两行代码:

```
e1 = Employee("Alex","alex@company_name.com","Finance",42,500000)
```

此语句创建类 Employee 的实例 e1。

```
e1.print_employee()
```

此语句调用 print_employee( )方法。通过参数 self 对 e1 进行访问,然后输出所有对象属性的值。输出结果如下。

```
Name :Alex
Email :alex@company_name.com
Department :Finance
Age :42
Salary :500000
```

现在,创建要输入表 2.1 中其余值的实例,并调用 print_employee( )方法输出该值。

```
class Employee:
    def __init__(self, name, email, department, age, salary):
        self.name    =   name
        self.email   =   email
        self.department  =  department
        self.age    =   age
        self.salary   =   salary

    def print_employee(self):
        print("Name  :  ",self.name)
        print("Email  :  ",self.email)
        print("Department  :  ",self.department)
```

```
        print("Age     : ",self.age)
        print("Salary  : ",self.salary)

#创建类实例 Employee 并调用 print_employee()方法
e1 = Employee("Alex","alex@company_name.com","Finance",42,500000)
e1.print_employee()

e2 = Employee("Evan","evan@company_name.com","HR",34,300000)
e2.print_employee()

e3 = Employee("Maria","maria@company_name.com","HR",30,350000)
e3.print_employee()

e4 = Employee("Pradeep","pradeep@company_name.com","IT",28,700000)
e4.print_employee()

e5 = Employee("Simon","simon@company_name.com","Finance",40,500000)
e5.print_employee()

e6 = Employee("Venkatesh","venkatesh@company_name.com","Sales",25,
200000)
e6.print_employee()
```

2.2 析构函数__del__()

学习 Python 类的__init__()函数后，读者自然想知道 Python 是否也有像其他语言（如 C++）中一样的析构函数。析构函数是在对象结束其生命周期时，系统自动执行的特殊函数。此方法的目的是释放对象占用的内存。在 Python 中，因为遵循用于内存管理的垃圾回收机制，所以析构函数需求较少，但它在 Python 中的确存在，并且实际中无须显式定义。析构函数是__del__()方法。未显式定义时会自动调用来销毁对象并清理对象占用的内存。

代码：

```
class Students:
    student_id = 50

    # 这是 Python 在创建类的新实例时调用的一个特殊函数
    def __init__(self, name, age):
        self.name = name
        self.age = age
        Students.student_id = Students.student_id + 1
        print('Student Created')
```

```python
    #这种方法统计学生总数
    def Total_no_of_students(self):
        print("Total Number of students in the school are : 
        " + str(Students.student_id))
    def __del__(self):
        print('object {} life cycle is over.'.format(self.name))

# 创建对象 stu1
stu1 = Students('Paris', 12)
stu1.Total_no_of_students()

# 销毁对象 stu1
del stu1

#检查对象是否仍然存在
stu1.Total_no_of_students()
```

输出：

```
Student Created
Total Number of students in the school are : 51
object Paris life cycle is over.
Traceback (most recent call last):
  File "F:\2020\input.py", line 23, in <module>
    stu1.Total_no_of_students()
NameError: name 'stu1' is not defined
>>>
```

2.3 类变量的类型

类有两种变量类型。

（1）实例变量。

（2）类变量或静态变量。

类变量和实例变量的定义如下：

```
class Class_name:
    class_variable_name = static_value

    def __init__(instance_variable_val):
        Instance_variable_name = instance_variable_val
```

目前，除了使用构造函数或 __init__() 函数创建实例属性，还可以使用一些常见方法创建实例属性，如 setGuarantee() 方法。

代码：

```
class Table:
    total_tables = 0

    def __init__(self, wood_used, year):
        self.wood_used = wood_used
        self.year = year
        Table.total_tables = Table.total_tables + 1
        print('Table made of {} wood has been created.'.format
        (self.wood_used))

    def setGuarantee(self,guarantee):
        self.guarantee = 25

t1 = Table('Ebony',2020)
```

在创建对象 t1 时，其只有 wood_used 与 year 两个属性。可使用 __dict__( ) 方法检验。

```
class Table:
    total_tables = 0

    def __init__(self, wood_used, year):
        self.wood_used = wood_used
        self.year = year
        Table.total_tables = Table.total_tables + 1
        print('Table made of {} wood has been created.'.format
        (self.wood_used))

    def setGuarantee(self,guarantee):
        self.guarantee = 25

t1 = Table('Ebony',2020)
print(t1.__dict__)
```

输出：

```
Table made of Ebony wood has been created.
{'wood_used': 'Ebony', 'year': 2020}
```

到目前为止，对象没有名为 guarantee 的属性。现调用 setGuarantee( ) 方法再次检验。

代码：

```
class Table:
    total_tables = 0
    def __init__(self, wood_used, year):
        self.wood_used = wood_used
        self.year = year
        Table.total_tables = Table.total_tables + 1
        print('Table made of {} wood has been created.'.format
        (self.wood_used))
```

```python
    def setGuarantee(self,guarantee):
        self.guarantee = guarantee

t1 = Table('Ebony',2020)
print("Attributes before calling setGaurantee() method.",t1.__dict__)
t1.setGuarantee(25)
print("Attributes after calling setGaurantee() method.",t1.__dict__)
```

输出:

```
Attributes before calling setGaurantee() method. {'wood_used': 'Ebony', 'year': 2020}
Attributes after calling setGaurantee() method. {'wood_used': 'Ebony', 'year': 2020, 'guarantee': 25}
>>>
```

也可以使用 getattr( ) 方法代替 dict( ) 方法。

代码:

```
print("Value of guarantee after calling setGaurantee() method.",getattr(t1,'guarantee'))
```

输出:

```
Value of guarantee after calling setGaurantee() method. 25
```

到目前为止，本书都是通过 __init__( ) 方法或其他实例方法创建属性，因此本书处理的所有属性实际上都是实例属性。对于每个实例，其属性各自拥有，相互独立，可通过如下方式访问：

objectName.attributeName

另外，类属性被类的所有实例共享，对所有对象来说，其值相同。其直接在类中定义，因此除通过类方法表示外，还可以通过 objectName.attributeName 或 className.attributeName 访问。类属性可以被实例属性覆盖。

因此，仔细观察 Table 类的代码，就会发现名为 total_tables 的属性，每创建一个 Table 对象，该属性就加 1。

代码:

```python
class Table:
    total_tables = 0

    def __init__(self, wood_used, year):
        self.wood_used = wood_used
        self.year = year
        Table.total_tables = Table.total_tables + 1
        print('Table No.{} : Table made of {} wood has been created.'.format(Table.total_tables,self.wood_used))

    def setGuarantee(self,guarantee):
        self.guarantee = guarantee
```

```
t1  =  Table('Ebony',2020)
t2  =  Table('Teak',2019)
t3  =  Table('Mangowood',2018)
```

输出：

```
Table No.1 : Table made of Ebony wood has been created.
Table No.2 : Table made of Teak wood has been created.
Table No.3 : Table made of Mango wood wood has been created.
>>>
```

表 2.2 所示为操作属性的常用方法。

表 2.2

说明	形式
创建属性	objectname.attr = value
判断属性是否存在	hasattr（obj, name）
读取对象属性	getattr（obj, name [, default]）
设置属性（不存在则先创建）	setattr（obj, name, value）
删除属性	delattr（obj, name）

2.4 继承

面向对象语言允许重用代码。继承就是将代码的可重用性提升到另一个层次的方式。继承中存在超类和子类。子类拥有超类中不存在的属性。假设为狗舍开发软件程序，因此应有 dog 类，其具有所有狗的共同特征。然而，谈到具体品种时，因为每个品种间均存在差异，所以可为每个品种创建类。这些类将继承 dog 类的共同特征，并在其基础上增添自身属性，使每个品种不同于其他品种。现在，逐步尝试创建一个类，然后创建其子类。通过如下示例更易理解其原理。

步骤 1：使用 class 关键字定义 dog 类。

代码如下所示。

```
class dog():
```

步骤 2：现已创建一个类，可为其创建方法。

本示例中创建一个能输出"I belong to a family of Dogs"的方法。

代码如下所示。

```
def family(self):
    print("I belong to a family of Dogs")
```

目前代码如下所示。

```
class dog():
    def family(self):
        print("I belong to a family of Dogs")
```

步骤3：创建类 dog 的对象。
代码如下所示。

```
c = dog()
```

步骤4：类的对象可通过"."调用 family( ) 方法。
代码如下所示。

```
c.family()
```

最终代码如下所示。

```
class dog():
    def family(self):
        print("I belong to family of Dogs")
c = dog()
c.family()
```

执行该程序，输出如下所示。

```
I belong to family of Dogs
```

下面继续学习继承。它在面向对象编程中被广泛应用。通过继承，可在不修改现有类的情况下创建新类。现有的类称为基类，继承的新类称为派生类。基类的特性可被派生类访问。代码如下所示。

```
class germanShepherd(dog):
    def breed(self):
        print("I am a German Shepherd")
```

germanShepherd 类的对象可调用 dog 类的方法。代码如下所示。

```
class dog():
    def family(self):
        print("I belong to family of Dogs")
class germanShepherd(dog):
    def breed(self):
        print("I am a German Shepherd")
c = germanShepherd()
c.family()
c.breed()
```

输出如下所示。

```
I belong to family of Dogs
I am a German Shepherd
```

查看先前代码，发现 germanShepherd 类的对象可用于调用类的方法。

关于继承需明确以下内容。

（1）继承可从类派生出任意数量的类。

在如下代码中创建另一派生类 husky。germanShepherd 与 husky 两个类均调用 dog 类的 family( )方法和其本身的 breed( )方法。

代码：

```
class dog():
    def family(self):
        print("I belong to family of Dogs")

class germanShepherd(dog):
    def breed(self):
        print("I am a German Shepherd")

class husky(dog):
    def breed(self):
        print("I am a husky")

g = germanShepherd()
g.family()
g.breed()
h = husky()
h.family()
h.breed()
```

输出：

```
I belong to family of Dogs
I am a German Shepherd
I belong to family of Dogs
I am a husky
```

（2）派生类可覆盖其基类的任意方法。

代码：

```
class dog():
    def family(self):
        print("I belong to family of Dogs")

class germanShepherd(dog):
    def breed(self):
```

```
            print("I am a German Shepherd")

class husky(dog):
    def breed(self):
        print("I am a husky")

    def family(self):
        print("I am class apart")

g = germanShepherd()
g.family()
g.breed()

h = husky()
h.family()
h.breed()
```

输出：

```
I belong to family of Dogs
I am a German Shepherd
I am class apart
I am a husky
```

（3）方法可调用与其同名的基类方法。

在如下代码中，husky 类有 family() 方法，它调用基类的 family() 方法，并在其中添加自身代码。

代码：

```
class dog():
    def family(self):
        print("I belong to family of Dogs")

class germanShepherd(dog):
    def breed(self):
        print("I am a German Shepherd")

class husky(dog): def
    breed(self):
        print("I am a husky")

    def family(self):
        super().family()

        print("but I am class apart")
```

```
g = germanShepherd()
g.family()
g.breed()

h = husky()
h.family()
h.breed()
```

输出：

```
I belong to family of Dogs
I am a German Shepherd
I belong to family of Dogs
but I am class apart
I am a husky
```

>
>
> **知识要点**
>
> - Python 类与对象。
> - Python 是面向对象的语言。
> - 对象或实例属于数据结构，由其类定义。
> - 类描述对象特征。
> - 类由属性、数据成员和可通过点操作符（.）访问的方法组成。
> - 类是 Python 的基本构建块。
> - 创建类的目的是定义创建该类实例所需的所有参数。
> - 其在逻辑上对与实体相关的所有功能进行分组。
> - 理论上可以以任何方式创建类，但编程时最好将一个类分配给一个现实实体。
> - 类中定义的函数也称为方法。
> - 方法是类中定义的函数。
> - 使用实例变量、类变量和方法定义对象。
> - 类的定义方法如下：
>
>   ```
>   class Class_name:
>       code(consisting of functions, variables etc)
>   ```
>
> - 类实例化通过函数实现。
>
>   ```
>   object_name = class_name()
>   ```
>
> - class_name()函数可以创建类的实例，并将其分配给局部变量 object_name。

- 通过调用类名并传递__init__( )函数所需参数来创建实例对象。
- Python 自带析构函数，无须显式定义。
- 析构函数称为__del__( )方法。
- 当__del__( )方法未显式定义时，会自动调用来销毁对象，以便释放被对象占用的内存资源。
- 类变量为可被类所有实例共享的变量。数据成员保存与类及其对象相关的数据。
- 实例变量定义于类方法内部，并只与类的当前实例关联。
- 可使用点操作符访问对象属性，如果访问类变量，则须使用带有类名的点操作符。
- 类对象支持属性引用和实例化。在属性引用的情况下，可通过以下方式引用属性：object_name. object_attribute。

小结

学习创建类与使用对象后，创建小项目，并对迄今所学的概念进行实际演练。这有助于巩固基础。下一章将介绍如何在 Python 中处理文件。

选择题

1. 创建 Table 类的新实例：

```
class Table:
    total_tables = 0
    def __init__(self, wood_used, year):
        self.wood_used = wood_used
        self.year = year
        Table.total_tables = Table.total_tables + 1
        print('Table made of {} wood has been created.'.format(self.wood_used))
```

a. t1 = Table('Ebony',2020)
b. t1 = Table()
c. t1 = __init__( )
d. t1 = __init__('Ebony',2020)

2. 下列选项中，哪一项是如下代码的输出结果？

```python
class Table:
    total_tables = 0
    def __init__(self, wood_used, year):
        self.wood_used = wood_used
        self.year = year
        Table.total_tables = Table.total_tables + 1
        print('Table made of {} wood has been created.'.format(self.wood_used))
    def __del__(self):
        print('Beautiful Table made of {} wood has been destroyed.'.format(self.wood_used))

t1 = Table('Ebony',2020)
```

a. Table made of Ebony wood has been created.
b. Beautiful Table made of Ebony wood has been destroyed.
c. Table made of Ebony wood has been created.
d. Beautiful Table made of Ebony wood has been destroyed.
e. Table made of Ebony wood has been created.

答：

1. a
2. d

1. 类的实例称为什么？

答：对象。

2. 简单解释 self。

答：

（1）与 Java 中的 this 或 C++ 中的指针类似。

（2）Python 中所有函数定义时第一个参数均为（'self'），函数调用时，不传递该参数。

（3）即使函数没有实参，仍需在函数定义中提及形参 self。

3. 简述类的组成。

答：

（1）class 关键字。

(2) 实例和类属性。

(3) self 关键字。

(4) __init__() 函数。

4. 什么方法类似 Java 中的构造函数，调用于对象实例化时，以初始化对象？

答：__init__()。

5. 什么是多重继承？

答：一个类派生于多个类称为多重继承。

6. A 是 B 的子类。如何从 A 中调用 B 的 __init__() 函数？

答：

通过以下两种方法。

(1) super().__init__()

(2) __init__(self)

7. 在 Python 中如何定义鸟与鹦鹉的关系？

答：继承。鹦鹉是鸟的子类。

8. 火车与窗户之间是什么关系？

答：组合。

9. 学生与学科之间是什么关系？

答：关联。

10. 学校与老师之间是什么关系？

答：组合。

11. 编写代码创建空类 Bank。

答：

```
class Bank:
    pass
```

12. 调用时名称以下划线开头、结尾的方法称为什么？

答：特殊方法。

13. 创建时对象如何调用特殊方法 __init__()？

答：特殊方法 __init__() 不可显式调用，创建类的对象时自动被调用。

14. 在类的方法中定义的变量称为什么？

答：实例变量。

15. 定义在类中但在方法外的变量称为什么？

答：类变量。

16. 什么对类中所有对象均相同？

答：类属性。

17. 什么是__init__( )方法的参数，不同对象中其值不同？

答：实例变量。

18. 如何判断哪些是类变量属性，哪些是实例变量属性？

答：类变量属性与实例变量属性的对比见表 2.3。

表 2.3

在类结构范围内定义	类变量
通过点操作符后跟类名访问	类变量
类中所有实例共享	类变量
属于其类本身	类变量
由实例拥有	实例变量
通常对每个实例其值相同	类变量
定义在类头正下方	类变量
不同实例中其值不同	实例变量
仅可通过对象名访问	实例变量

编程题

1. 写出如下代码的输出。

```
class Kite:
    def fly(self):
        print('is meant to fly')
k = Kite()
k.fly()
```

答：

```
is meant to fly
```

2. 写出如下代码的输出。

```
class classLevel:
    def class1(self):
        print('There are 20 students in this class')
    def class2(self):
        print('There are 15 students in this class')
    def class3(self):
        print('There are 41 students in this class')
    def class4(self):
        print('There are 30 students in this class')
```

```
cl = classLevel()
cl.class3()
```

答：There are 41 students in this class。

3. 写出如下代码的输出。

```
class Twice_multiply:
    def __init__(self):
        self.calculate(500)

    def calculate(self,num):
        self.num = 2 * num;

class Thrice_multiply(Twice_multiply):
    def __init__(self):
        super().__init__()
        print("num_from_Thrice_multiply is",self.num)

    def calculate(self,num):
        self.num = 3 * num;

tm = Thrice_multiply()
```

答：

num from Thrice_multiply is 1500
>>>

4. 对于如下代码，能否验证 tm 为 Thrice_multiply 类的对象？

```
class Twice_multiply:
    def __init__(self):
        self.calculate(500)

    def calculate(self, num):
        self.num = 2 * num;

class Thrice_multiply(Twice_multiply):
    def __init__(self):
        super().__init__()
        print("num from Thrice_multiply is", self.num)

    def calculate(self, num):
        self.num = 3 * num;

tm = Thrice_multiply()
```

答：可以使用 isinstance( ) 函数验证实例是否属于一个类。

5. 写出如下代码的输出。

```
class Table:
    def features(x):
        print('This table has {} legs.' .format(x))
Table.features(4)
```

答：This table has 4 legs。

1. 阐述有界方法、无界方法与静态方法之间的区别。

答：有界方法的第 1 个参数是 self。它依赖实例，仅通过实例访问。

无界方法不将 self 作为参数，并且从 Python 开始不再使用。

静态方法绑定到类而非对象，因此通过类名访问。

在无须修改实例的情况下，静态方法用作实用函数。例如，求数的阶乘可以使用 maths 类。

```
>>>math.factorial(8)
```

求阶乘不必创建对象来调用函数，可以使用@ staticmethod 装饰器实现静态方法。

代码：

```
class classLevel:
    def class1(self,string_a):
        print(string_a)

    def class2(self):
        print('There are 15 students in this class')

cl = classLevel("Hello")
cl.class1()
cl.class2()
```

输出：

```
Traceback (most recent call last):
  File "F:\2020 - BPB\Python for Undergraduates\code\input.py",
  line 8, in <module>
    cl = classLevel("Hello")
TypeError: classLevel() takes no arguments
>>>
```

使用静态装饰器的方法如下。

代码:

```
class Foo:
    @staticmethod
    def bar():
        print('Static method bar()')

    @staticmethod
    def stat():
        print("Static method stat()")

    def class1(self,string_a):
        print(string_a)
#调用静态方法 bar()
Foo.bar()

#调用静态方法 stat()
Foo.stat()

#调用有界方法 class1()
f = Foo()

#调用有界方法 class1()
f = Foo()
f.class1('bounded method')
```

输出:

```
Static method bar()
Static method stat()
bounded method
>>>
```

2. 什么是魔术方法?

答: 魔术方法也称为特殊方法或 dunder (表示双下划线) 方法。因其不被直接调用,加、减等每个运算符都拥有其自身相关的魔术方法。Python 定义了诸多魔术方法。示例如下。

(1) __init__():初始化对象。

(2) __add__():使用"+"调用此方法。

(3) __str__():输出对象是自动调用。

(4) __lt__():使用"<"调用此方法。

魔术方法的调用都在幕后。因此,在输入"2+3"时,add()方法即被调用。魔术方法最大的优点在于使对象行为类似内置类型。

当"+"操作符用于两个整数值 10 和 10 时,返回的整数值为 20。当"+"操作符用于两个字符串 10 和 10 时,返回的字符串值为 1010。这是因为"+"为自动调用的魔术方法 add()的缩写。当"+"的左边操作数是数字时,"+"

作为加法操作符；当"+"的左边操作数是字符串时，则"+"作为字符串连接操作符。

```
>>>10 +10
20
>>> '10'+'10'
'1010'
>>>
```

3. 什么是运算符过载？

答：将多个函数赋值给某个特定操作符称为运算符过载。读者现已学过魔术方法的原理，假设你和朋友在同一银行均有一个银行账户。

代码：

```
class Bank:
    def __init__(self, name, balance):
        self.name = name
        self.balance = balance

b1 = Bank('customer1','10000')
b2 = Bank('friend','10000')

#打印 b1 和 b2 的值
print(b1)
print(b2)

#打印 b1 与 b2 的和值
b1 + b2
```

假设输出两个对象的值，将得到对象的内存位置而非数据。此外，如果试图将两个对象相加以求总余额，则会报错，因为未定义两个对象的添加。

输出：

```
<__main__.Bank object at 0x03228E20 >
<__main__.Bank object at 0x034AB3D0 >
Traceback (most recent call last):
  File "F:\2020 - BPB\Python for Undergraduates\code\input.py",
  line 15, in <module>
    b1 +b2
TypeError: unsupported operand type(s) for +:'Bank' and 'Bank'
```

可以通过过载__str__( )方法和__add__( )方法定义操作符功能。

代码：

```python
class Bank:

    def __init__(self.name,balance):
        self.name = name
        self.balance = balance
    def __str__(self):
        str1 ='name:'+self.name +'balance:'+self.balance +'.'
        return str1
    def __add__(self, other):
        balance = int(self.balance) + int(other.balance)
        return balance

b1 = Bank('customer1','10000')
b2 = Bank('friend','10000')

#打印 b1 和 b2 的值
print(b1)
print(b2)

#打印 b1 与 b2 的和值
print(b1 + b2)
```

输出：

```
name: customer1 balance: 10000.
name: friend balance: 10000.
20000
>>>
```

第 3 章 文 件

引言

文件是生活中常见的名称。你的父亲可能会把银行文件、办公室文件、汽车文件等分开保存。同样，你也会将学业中的不同项目保存至不同文件。因此，人们基本上都需要文件来存储重要信息，这同样适用于计算机。将重要数据存储在计算机文件中。这时，使用数据庞大的硬拷贝并不明智，因此可将所有需要信息存储在系统文件中。

知识结构

- 文件存储的优点
- 目录与文件管理
- 使用 getcwd() 方法读取当前工作目录
- 使用 listdir() 方法读取目录内容
- 使用 mkdir() 方法创建目录
- 使用 rename() 方法重命名目录
- 使用 rmdir() 方法删除目录
- 文件操作
    - open() 函数与 close() 函数
    - 文件写入的访问模式
    - 使用 wb 访问模式写入二进制文件
    - 使用 rb 访问模式读取二进制文件内容
    - 使用二进制文件读写字符串
    - 其他二进制文件操作
    - 文件属性
- 各种文件操作方法

第 3 章 文件

> **目标**
> 完成本章的学习后,读者应掌握以下技能。
> (1) 使用 Python 管理目录和文件。
> (2) 检索当前工作目录下的目录。
> (3) 检查 Python 目录内容。
> (4) 使用 mkdir( )方法创建个人目录。
> (5) 使用 rename( )方法重命名目录。
> (6) 使用 rmdir( )方法删除目录。
> (7) 处理文件。

文件是一组存储数据的连续字节。数据可存储于".txt"文件或".exe"文件中。计算机在处理文件时,必须将数据转换成二进制格式。下面进一步学习文件的相关知识。

3.1 文件存储的优点

在使用文件之前,应先了解文件存储的优点。文件存储的优点如下。
(1) 数据永久保存在文件中,即使关闭计算机,其仍保存在系统中。
(2) 可随时轻松更新文件。
(3) 存储在计算机中的数据可用于各种复杂计算。
(4) 可无限存储数据,这在物理学中是无法实现的。此外,数据存储于文件中,既可用于分析,也可进行搜索。

3.2 目录与文件管理

文件包含头部、数据和文件结尾 3 个主要部分,如图 3.1 所示。

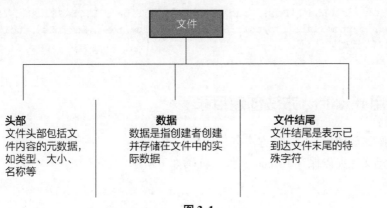

图 3.1

虽然可处理的文件类型不限,但目前仅处理 ".txt" 文件和 ".csv" 文件。

在处理文件之前,首先学习目录与文件管理。在计算机中,文件组织于一个目录。如果使用该目录,则需导入 os 模块。

```
import os
```

3.3 使用 getcwd( )方法读取当前工作目录

如果检索当前工作目录(cwd)的绝对路径,则需导入 os 模块并调用 getcwd( )方法。代码如下所示。

```
>>> import os
>>> os.getcwd()
'C:\\Users\\MYPC\\AppData\\Local\\Programs\\Python\\Python38-32'
```

通过 print( )方法输出当前工作目录名。代码如下所示。

```
>>> import os
>>> print(os.getcwd())
C:\Users\MYPC\AppData\Local\Programs\Python\Python38-32
```

结合转义字符可知,在输出('\\')时将转义为 \ 。

3.4 使用 listdir( )方法读取目录内容

可以使用 listdir( )方法读取目录内容,并以 Python 列表形式显示目录中的所有内容。

```
>>> import os
>>> os.listdir()
['comment.py','DLLs','Doc','include','Lib','libs','LICENSE.txt','NEWS.txt',
'python.exe','python3.dll','python38.dll','pythonw.exe','Scripts','tcl','Tools',
'vcruntime140.dll']
```

3.5 使用 mkdir( )方法创建目录

本节介绍如何使用 mkdir( )方法创建目录。
首先创建文件夹以保存 Python 文件。代码如下所示。

```
>>> import os
>>> os.mkdir("Learning Python")
```

调用 listdir( )方法，将显示新文件夹名称。代码如下所示。

```
>>>os.listdir()
['comment.py','DLLs','Doc','include','Learning Python','Lib','libs',
'LICENSE.txt','NEWS.txt','python.exe','python3.dll','python38.dll',
'pythonw.exe','Scripts','tcl','Tools','vcruntime140.dll']
```

也可进入文件资源管理器。如果在其内访问相同路径（本示例中为 C：\Users\MYPC\AppData\Local\Programs\Python\ Python38 - 32），可看到新建文件夹，如图 3.2 所示。

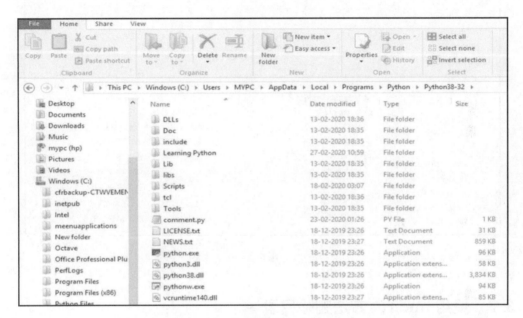

图 3.2

3.6 使用 rename( )方法重命名目录

可以使用 rename( )方法将文件夹"Learning Python"重命名为"Understanding Python"，如图 3.3 所示。代码如下所示。

```
>>> import os
>>>os.rename("Learning Python","Understanding Python")
>>>os.listdir()
['comment.py','DLLs','Doc','include','Lib','libs','LICENSE.txt','NEWS.txt',
'python.exe', 'python3.dll', 'python38.dll', 'pythonw.exe', 'Scripts', 'tcl',
'Tools','Understanding Python','vcruntime140.dll']
```

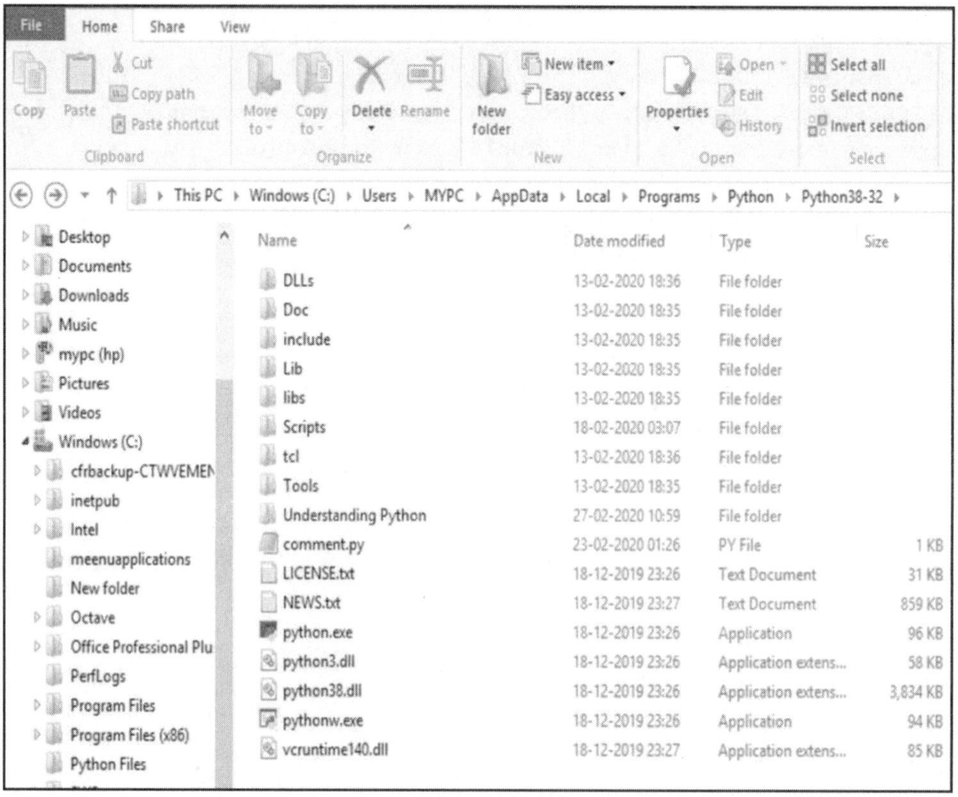

图 3.3

3.7 使用 rmdir( )方法删除目录

可以使用 rmdir( )方法删除目录。

在同一目录下创建名为"delete_ file. txt"的新文本文件,如图 3.4 所示。

现在,介绍如何从目录中删除此文件。代码如下所示。

```
>>> import os
>>>os.remove("delete_file.txt")
```

文件不存在。调用 os. path. exists( )函数检验目录是否存在。代码如下所示。

```
>>> import os
>>>os.path.exists('C:\\Users\\MYPC\\AppData\\Local\\Programs\\Python\\Python38-32')
True
```

删除文件后如图 3.5 所示。

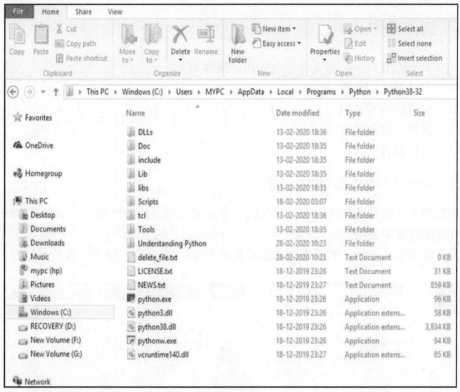

图 3.4

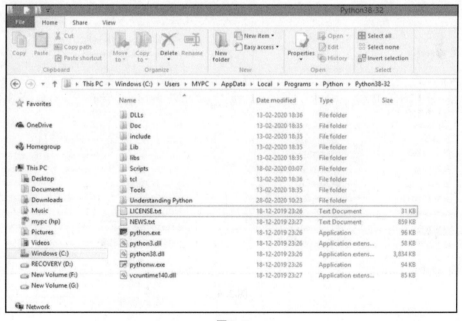

图 3.5

3.8 文件操作

本节学习使用 Python 对文件进行操作。具体如下。
(1) 打开与关闭文件。
(2) 文件访问模式。
(3) 使用二进制文件。
(4) 文件属性。

3.8.1 open()函数与close()函数

在处理文件前需先打开文件,此为第一步。完成对文件的操作后,要将其关闭。所有已打开文件必须关闭。

为了演示文件操作,需要先创建文本文件"first_file.txt",如图3.6所示。

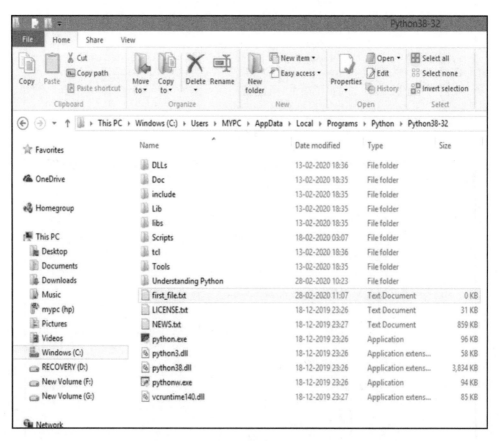

图 3.6

文件内容如图 3.7 所示。

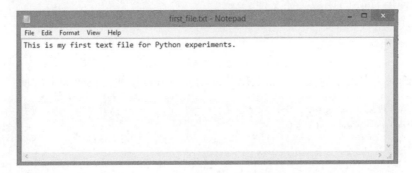

图 3.7

(1) 使用 open() 函数打开文件。

语法:

```
file_obj = open(file_name, access_mode, buffering)
```

① file_name: 要访问的文件名称, 为强制参数。

② access_mode: 可选参数, 默认模式为只读 (r)。其用于指定文件访问模式。表 3.1 所示为各种类型的访问模式。

③ buffering: 可选参数, 1 表示使用指定大小的缓冲区; 小于 1 表示使用默认大小的缓冲区。

```
file_handler = open("first_file.txt")
```

在操作文件前, 应学习如何关闭文件。任何已打开文件必须关闭, 以避免意外发生。在执行程序时, 已打开文件可能引发错误。因此, 为保险起见, 文件完成操作后必须关闭。

表 3.1

访问模式	名称	说明
r	读取	文件以只读模式打开, 文件指针置于文件开头。为默认模式
r +	读写当前文件	以此访问模式打开文件用于读写。文件指针置于文件开头
w	写入	以此访问模式打开文件只用于写入。如果文件存在, 则将其覆盖; 如果文件不存在, 则创建新文件, 并将内容写入
w +	读写	以此访问模式打开文件用于读写。如果文件不存在, 则新建一个同样名字的文件, 并对文件进行读写
a	追加	文件末尾追加内容。如果文件不存在, 则创建新文件将内容写入
a +	读取与追加	以追加模式打开文件用于读写, 如果文件不存在, 则将创建该文件

（2）使用 close( )函数关闭文件。

语法：

```
file_handler.close()
```

为了确保错误发生时文件正常关闭，使用 try…finally 块。代码如下所示。

```
file_handler = open("first_file.txt")
try:
        #代码
        ……
finally:
        file_handler.close()
```

或者使用 with 关键字。代码如下所示。

```
>>> with open("first_file.txt") as file_handler:
```

3.8.2 文件写入的访问模式

现在操作文件"first_file.txt"。

代码：

```
#使用 read( )函数以字符串形式提取文件的所有内容
>>> file_handle = open("first_file.txt","r")
>>> try:
    file_handle.read()
finally:
    file_handle.close()
```

输出：

```
"This is my first text file for Python experiments.\nThere is a lot that can be done with files in Python.\nLet's get Started."
```

或

```
#使用 read( )函数提取文件前端部分字符（本例为 10 个）
>>>try:
    file_handle.read(10)
finally:
    file_handle.close()
```

输出：

```
'This is my'
```

或

```
#逐行打印
>>>file_handle = open("first_file.txt","r")
>>>try:
    for each in file_handle:
        print(each)
finally:
    file_handle.close()
```

输出：

```
This is my first text file for Python experiments.
There is a lot that can be done with files in Python.
Let's get Started.
```

(1) 以 r+ 模式读取文件。

代码：

```
>>>file_handle = open("first_file.txt","r+")
>>>try:
    file_handle.read()
finally:
    file_handle.close()
```

输出：

```
"This is my first text file for Python experiments.\nThere is a lot that can be done with files in Python.\nLet's get Started.\n"
```

(2) 以 r+ 模式读写文件。

代码：

```
file_handle = open("first_file.txt",'r+')
>>>try:
    file_handle.write("Now I am using r+ mode")
    file_handle.read()
finally:
    file_handle.close()
```

输出：

```
22
"file for Python experiments.\nThere is a lot that can be done with files in Python.\nLet's get Started.\n"
```

打开该文件，内容如图 3.8 所示。

```
Now I am using r+ modefile for Python experiments.
There is a lot that can be done with files in Python.
Let's get Started.
```

图 3.8

数字22表示从22号位读取。该处共插入22个字符。因此，使用r+模式将插入内容置于文件开头。然而，有时存在其他需求。如果在任意位置插入文本，则需使用seek()函数。seek()函数有两个参数：第1个参数为读/写指针在文件中的位置；第2个参数为可选参数，默认值为0，表示文件开头。如果在当前位置插入文本，则第2个参数值应改为2。

（3）在文件末尾插入文本。代码如下所示。

```
>>>file_handle = open("first_file.txt",'r+')
>>>try:
file_handle.seek(0,2)
file_handle.write("\n Now I am using r + mode. Inserting at the end of file")
file_handle.read()
finally:
    file_handle.close()
```

如果使用r+模式，则文件需存在于系统中，如图3.9所示；否则，将会报错，如图3.10所示。

```
This is my first text file for Python experiments.
There is a lot that can be done with files in Python.
Let's get Started.
    Now I am using r+ mode. Inserting at the end of file
```

图 3.9

```
>>> file_handle = open("my_file.txt",'r+')
Traceback (most recent call last):
  File "<pyshell#56>", line 1, in <module>
    file_handle = open("my_file.txt",'r+')
FileNotFoundError: [Errno 2] No such file or directory: 'my_file.txt'
>>>
```

图 3.10

```
>>>file_handle = open("first_file.txt","w")
>>> try:
    file_handle.write("Hi There")
finally:
    file_handle.close()
```

以上代码将"Hi There"写入文件，如图3.11所示。

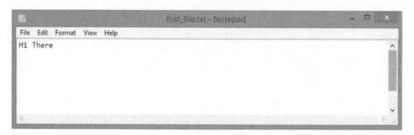

图 3.11

如果该文件不存在，则按给定名称创建文件，并添加内容。代码如下所示。

```
>>>file_handle = open("first_file1.0.txt","w")
>>>try:
    file_handle.write("Hi There")
finally:
    file_handle.close()
```

由于不存在名为"first_file1.0.txt"的文件，所以创建该文件，如图 3.12 所示。

图 3.12

文件内容如图 3.13 所示。

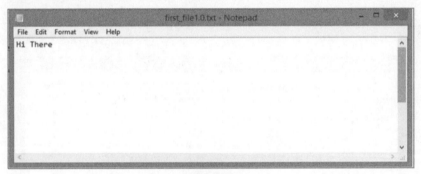

图 3.13

访问模式为 w 时，不执行读操作。代码如下所示。

```
>>>file_handle = open("first_file.txt","w")
>>>try:
    file_handle.read()
finally:
    file_handle.close()

Traceback (most recent call last):
  File "<pyshell#17 >", line 2, in <module>
file_handle.read()
io.UnsupportedOperation: not readable
```

访问模式为 w+ 时，如图 3.14 所示。代码如下所示。

```
>>>file_handle = open("first_file.txt","w+")
>>> try:
    file_handle.write("Hi")
finally:
    file_handle.close()
```

图 3. 14

访问模式为 a 时，如图 3.15 所示。代码如下所示。

```
>>>file_handle = open("first_file.txt","a")
>>>try:
    file_handle.write("\n This is so exciting.")
finally:
    file_handle.close()
```

图 3. 15

3.8.3　使用 wb 访问模式写入二进制文件

如下代码演示如何使用 wb 访问模式写入二进制文件。

```
#打开文件时,如果该名称文件不存在,则创建该文件,如图 3.16 所示
>>>f =open("my_bin_file.bin", "wb")
>>>num =[1,2,3,4,5]
```

图 3. 16

```
# 如果要存储列表,则需先将其转换为字节数组,以获得可添加到二进制文件的字节表示
>>>arr  =bytearray(num)

#只检查字节表示的样子
```

```
>>>print(arr)
bytearray(b'\x01\x02\x03\x04\x05')
>>>f.write(arr)
5
>>>f.close()
```

3.8.4 使用 rb 访问模式读取二进制文件内容

可以使用 rb 访问模式读取二进制文件内容。

```
>>> f = open("my_bin_file.bin", "rb")
>>> number = list(f.read())
>>> print(number)
[1, 2, 3, 4, 5]
```

3.8.5 使用二进制文件读写字符串

如下代码演示如何使用二进制文件读写字符串。

代码：

```
#写入
file_handle = open("my_bin_file.bin", "wb")
msg = "Hi there"
arr = bytearray(msg.encode())
file_handle.write(arr)
file_handle.close()

#读取
file_handle = open("my_bin_file.bin","rb")
msg = file_handle.read()
print(msg)
print(msg.decode())
```

输出：

```
b'Hi there'
Hi there
```

3.8.6 其他二进制文件操作

表 3.2 所示为其他二进文件访问模式。

表 3.2

访问模式	名称	说明
rb +	对现有文件进行读写	对现有二进制文件执行读写操作

续表

访问模式	名称	说明
wb +	读写	对二进制文件进行读写操作。如果该文件不存在，则创建一个同名新文件
ab	追加	将内容追加至文件末尾
ab +	读取与追加	读取文件并于文件末尾追加新内容。如果该文件不存在，则创建一个同名新文件

3.8.7 文件属性

如下代码演示如何使用文件属性查询文件打开状态，以及文件模式和名称。

代码：

```
file_handle = open("first_file.txt","w+")
print('Closed or Not - ',file_handle.closed)
print('Mode :',file_handle.mode)
print('Name :',file_handle.name)
```

输出：

```
Closed or Not -  False
Mode :  w+
Name :  first_file.txt
```

3.9 各种操作文件的方法

表 3.3 所示为各种操作文件的方法。

表 3.3

方法	说明
close( )	关闭文件
flush( )	刷新内部缓冲区内存
fileno( )	返回整数
next( )	返回文件下一行
read( )	读取文件
readline( )	读取整行
readlines( )	逐行返回直到 ie 结束，然后返回文件行列表
seek( )	更改位置

续表

方法	说明
tell( )	返回文件指针当前位置
truncate( )	截断文件
write( )	写入文件
writelines( )	将字符串列表写入文件

示例 3.1

readline( )方法,代码如下所示。

```
>>>file_handle = open("first_file.txt","r")
>>>  try:
    file_handle.readline()
finally:
    file_handle.close()

'This is my first text file for Python experiments.\n'
```

示例 3.2

readlines( )方法, 代码如下所示。

```
File_handle = open("first_file.txt","r")
>>> try:
    file_handle.readlines()
finally:
    file_handle.close()

['This is my first text file for Python experiments.\n','There is a lot that can be done with files in Python.\n', "Let's get Started."]
```

示例 3.3

writelines( )方法, 代码如下所示。

```
>>>file_handle = open("first_file.txt","w")
>>>  try:
file_handle.writelines(['Hi \n','How are you \n','Good to see you :)'])
finally:
file_handle.close()
```

本章介绍了如何用文件存储数据。数据也能以表格形式存储在数据库中。下一章将介绍 MySQL 数据库,以及如何使用 Python 访问数据库和执行所有类型的 SQL 查询。

选择题

1. 下列选项中，允许对现有二进制文件执行读写操作的是（　　）。
   a. rb +　　　　b. wb +　　　　c. ab　　　　d. ab +
2. 下列选项中，允许对二进制文件进行读写操作，如果该文件不存在，则创建同名新文件的操作是（　　）。
   a. rb +　　　　b. wb +　　　　c. ab　　　　d. ab +
3. 下列选项中，允许将内容附加至文件末尾的操作是（　　）。
   a. rb +　　　　b. wb +　　　　c. ab　　　　d. ab +
4. 下列选项中，允许读取文件并于文件末尾追加新内容，如果该文件不存在，则创建同名新文件的操作是（　　）。
   a. rb +　　　　b. wb +　　　　c. ab　　　　d. ab +
5. 下列选项中，哪一项是如下代码的功能？（　　）

   ```
   file_handle = open("my_bin_file.bin","rb")
   ```

   a. 打开文件，仅供写入
   b. 打开文件，用于读写
   c. 以二进制格式打开文件，用于读写
   d. 以二进制格式打开文件，仅用于读取

答
1. a
2. b
3. c
4. d
5. d

简答题

1. 用于删除文件的函数是什么？

答：remove( )函数。

```
>>> import os
>>>os.remove("fileName.txt")
```

2. 如何在当前目录中创建目录。

答：

```
>>> import os
>>>os.mkdir("directory_name")
```

3. chdir( )方法有何用途？

答：chdir( )方法用于更改当前目录。其接收一个参数，作为当前目录更改后的目录名。

```
>>> import os
>>> os.chdir('directory_name')
```

4. 哪种方法用于读取当前工作目录？

答：getcwd( )方法。

```
>>> import os
>>> os.getcwd()
```

5. rmdir( )方法的功能是什么？

答：rmdir( )方法用于删除目录。它将目录名作为参数传递。

填空题

1. _____ 用于永久大量存储数据和文件集合。
2. 文件分两种方式存储：_____ 和 _____ 。
3. _____ 指定打开文件的操作类型。
4. _____ 文件以 ASCII 或 Unicode 字符存储信息。
5. 在文本文件中，每行文本都由特殊字符 _____ 终止。
6. _____ 文件将信息以其存储在内存中的格式原样进行存储。
7. 在二进制文件中，每行均无 _____ 。
8. "fileObject._____（sequence_of_strings）"用于将字符串写入文件。
9. _____ 方法用于读取整个文件并以字符串形式返回文本。
10. _____ 方法用于查找文件当前位置。
11. 文件对象的 _____ 方法将清除所有未写入信息并关闭文件对象。
12. tell( )方法返回 _____ 值，其用于读取文件指针的当前位置。

答：

1. 文件
2. 文本文件，二进制文件

3. 访问模式

4. 文本

5. End of line(EOL)

6. 二进制

7. 分隔符

8. writelines

9. read( )

10. tell( )

11. close( )

12. 整数

13. 查看如下代码，并回答以下问题。

```
fileHandle = open('x','a')
/_____
fileHandle()
```

a. x 为何种文件类型？
b. 在空白处填写将"Hello World"写入文件 x 的语句。

答：

a. x 为文本文件
b. fileHandle.write ('Hello World')

编程题

1. 将文件夹的名称由 mar2020 更改为 jun2020。

答：

```
>>> import os
>>> os.rename("mar2020","jun2020")
```

2. 使用 Python 编写一条语句执行以下操作。

a. 以 w 模式打开文件"MYPET.txt"。
b. 以 r 模式打开文件"MYPET.txt"。

答：

a. file_handle = open("MYPET.txt","w")
b. file_handle = open("MYPET.txt","r")

问答题

1. Python 中的文件访问模式 r+ 和 w+ 有何不同？

答：

（1）w+ 表示读取和写入操作。此模式以读写模式打开文件，如果文件不存在，则创建同名新文件用于读写。

（2）r+ 表示对现有文件进行读写。此模式将打开文件用于读写。文件指针置于文件开头。

2. Python 中的文件访问模式 r+ 和 rb+ 有何异同？

答：

（1）r+ 表示对现有文件进行读写。此模式将打开文件用于读写。文件指针置于文件开头。

（2）rb+ 表示对现有文件进行读写（对现有二进制文件执行读写操作）。

3. 如何使用 with 语句？

答：with 语句用于编写两个互为关联的语句。

```
'with open(filename,mode) as fileHandle:
----- code to be execute -----
```

即使代码结束之前出现异常，该语句也会在嵌套代码块后自动关闭文件。

第 4 章

MySQL与Python交互

第 4 章　MySQL 与 Python 交互

引言　软件应用通常与存储信息的数据库关联。本章介绍如何安装 MySQL 数据库，使用 Python 对其进行配置，并使用 Python 与 MySQL 交互。

知识结构
- 安装和配置 MySQL
  - 安装 MySQL
  - 配置 MySQL
- 使用命令行工具在 MySQL 中创建数据库
- 连接 MySQL 数据库与 Python
  - 使用 Python 在 MySQL 中创建数据库
  - 使用 Python 检索数据库记录
- 创建数据库
  - 使用 MySQL 创建数据库
  - 使用 Python 创建数据库
- 使用 Python 操作数据库
  - 使用 INSERT 语句插入记录
  - 读取记录
  - 使用 WHERE 子句读取表中选定记录
  - 使用 ORDER BY 子句对查询结果进行排序
  - 使用 DELETE 指令删除记录
  - 使用 UPDATE 指令更新记录

完成本章学习后，读者应掌握以下技能。
（1）在计算机中安装和配置 MySQL。
（2）执行 SQL 查询。
（3）使用 Python 创建 MySQL 数据库。
（4）使用 Python 在 MySQL 数据库上执行所有类型的 SQL 查询。
本章从学习安装 MySQL 与复习 SQL 基本概念开始，之后学习使用 Python 执行 SQL 查询。

4.1 安装和配置 MySQL

除了 Python 基础知识外，使用 MySQL 理解结构化查询语言（SQL）同样重要。SQL 是为存储、创建、更新、管理与删除数据而设计的标准语言。本节介绍如何安装和配置 MySQL。

4.1.1 安装 MySQL

如果服务器中已安装了 MySQL，可以跳过本小节。本小节将演示在 Windows 计算机系统中安装 MySQL 6.0 所需的步骤。此外，MySQL Essential 也必不可少，其对于初学者来说优点很多。如果在独立系统中学习，建议初学者通过网址 https://mysql-essential.en.uptodown.com/windows 安装 MySQL Essential。可遵循以下步骤。

步骤 1：下载 MySQL Essential（https://mysql-essential.en.uptodown.com/windows），如图 4.1 所示。

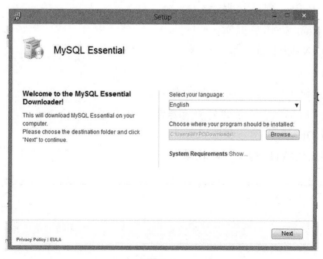

图 4.1

步骤 2：双击 mysql – essential – 6.0.0. msi 软件包以启动安装。
（1）显示 MySQL Server6.0 安装向导窗口。
（2）如果要继续执行安装过程，则单击 "Next" 按钮。
（3）选择设置类型。共 3 种类型。
①Typical（默认安装）包含常用功能。
②Complete（完全安装）包含所有程序功能。
③Custom（用户自定义安装）自定义选择所有安装程序。建议选择 Typical 类型安装。如果想更改软件安装路径，可选择 Custom 类型安装。
（4）本书选择 Typical 类型。
（5）下一个窗口提示登录或创建 MySQL 账户，可以跳过注册。
（6）可以创建账户或单击 "Skip Sign – Up" 按钮，然后单击 "Next" 按钮。
（7）为安装过程的最后窗口。
（8）在单击 "Finish" 按钮前，务必勾选 "Configure the MySql Server now" 复选框。

4.1.2 配置 MySQL

配置 MySQL 的操作步骤如下。
（1）单击 "Finish" 按钮后，弹出 "MySQL Server Instance Configuration Wizard（MySQL 服务器实例配置向导）" 窗口。
（2）如果该窗口未自动弹出，则单击桌面上的 "开始" 按钮，选择 "Wizard（向导）" 选项。在第一个对话框中单击 "Next" 按钮。按以下步骤操作。
①选择详细配置或标准配置。本书选择 Detailed Configuration（详细配置）。
②选择服务器类型。该选择将影响内存、磁盘和 CPU 的使用情况。如果在承载其他应用程序的服务器上工作，则选择 Server（服务器），然后单击 "Next" 按钮。
③如果上一步中选择 Server，则会显示要求为 InnoDB 数据文件设置路径，以启用 InnoDB 数据库引擎。此处不做修改并单击 "Next" 按钮。
④设置到服务器的大约并发连接数。在理想情况下，若无特殊需求则推荐选择第一个选项，即 "Decision Support（DSS）/OLAP（决策支持）"。完成选择后单击 "Next" 按钮。
⑤在下一个对话框中，存在两个选项：启用 TCP/IP 组网、启用严格模式。查看 "启用 TCP/IP 组网" 选项。默认选择第二选项 "启用严格模式"，但该选项不适合大多数应用程序，因此，如果没有充分理由选择 "启用严格模式" 选项，则取消选择该选项，并单击 "Next" 按钮。
⑥设置 MySQL 默认字符集。选择标准字符集，然后单击 "Next" 按钮。
⑦设置窗口选项。可见到 3 个复选框："Install as Windows Service" "Launch the MySQL Server automatically" 和 "Include Bin Directory in Windows PATH"。3 个复选框均勾选后，单击 "Next" 按钮。
⑧设置账户 root 密码。勾选 "Modify Security Settings（修改安全设置）" 复选框并

提供密码详细信息。如果服务器在互联网中,则请勿勾选"Enable root access from remote machines(启用远程机器的根访问)"复选框。另外,不建议勾选"Create An Anonymous Account(创建匿名账户)"复选框。请将密码保存于安全地方。

⑨此窗口显示配置处理。处理结束后,单击"Finish"按钮,即可完成安装,如图 4.2 所示。

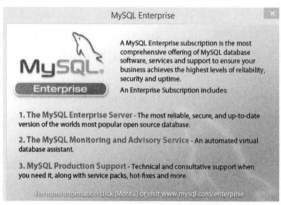

图 4.2

注意:查找 MySQL 信息时,可能遇到 MySQL community server、MySQL installer 与 MySQL Essentials 等术语。

顾名思义,Essentials 只提供基础功能,无额外组件和模板。installer 和 community server 具有完整服务器特性,不同的是 community server 为在线安装,而 installer 需下载安装包,并且可离线安装。

4.2 使用命令行工具在 MySQL 中创建数据库

现在已经完成 MySQL 配置,打开命令提示符,并在框中输入以下指令。

```
mysql -u root -p
```

完成后将提示输入密码,输入密码后,命令提示符如图 4.3 所示。

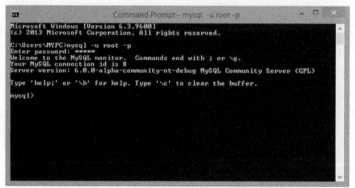

图 4.3

学习 MySQL，首先要创建数据库。为此可遵循以下步骤。
步骤 1：启动命令提示符窗口。
按"Win + R"组合键后输入"cmd"，或者单击"开始"按钮，搜索"cmd"。命令提示符窗口将被打开。
步骤 2：登录账户。
输入如下指令。

```
mysql - u root - p
```

按 Enter 键，命令提示符窗口中显示"Enter password（输入密码）"。
（1）输入配置时设置的 root 密码。
（2）创建数据库用于管理工作，并将其命名为"PYMYSQL"。
（3）输入如下指令。
语法：

```
CREATE DATABASE DATABASE_NAME;
```

假设为 Python 程序创建一个名为"PYMYSQL"的 MySQL 数据库。其创建指令如下：

```
CREATE DATABASE PYMYSQL;
```

学习 MySQL 还要了解其查询的基本规则。
（1）MySQL 命令不区分大、小写。
（2）数据库名称不能含有空格。
（3）所有 MySQL 命令均以分号（;）结尾。如果忘记输入分号，则可在下一行输入并按 Enter 键，指令仍可执行。
如果数据库创建成功，则出现如下消息。

```
Query OK, 1 row affected
```

4.3 连接 MySQL 数据库与 Python

Python 数据库访问模块的标准接口由 Python 数据库 API 定义。所有 Python 数据库模块均遵循此接口。
这有助于实现接口一致性。由于涉及过程类似，所以该 API 更易处理不同数据库。以下为处理数据库时需遵循的基本步骤。该 API 包括以下内容。
（1）导入 API 模块。
（2）连接数据库。
（3）执行 SQL 查询。
（4）断开连接。
使用指令 pip install mysql – connector – python 安装 MySQL 连接器。

4.3.1 使用 Python 在 MySQL 中创建数据库

在命令提示符窗口中输入以下指令（如图 4.4 所示）。

```
show databases;
```

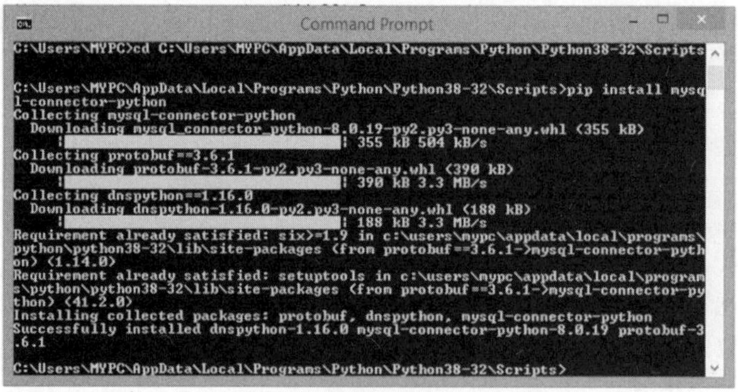

图 4.4

显示数据库如图 4.5 所示。

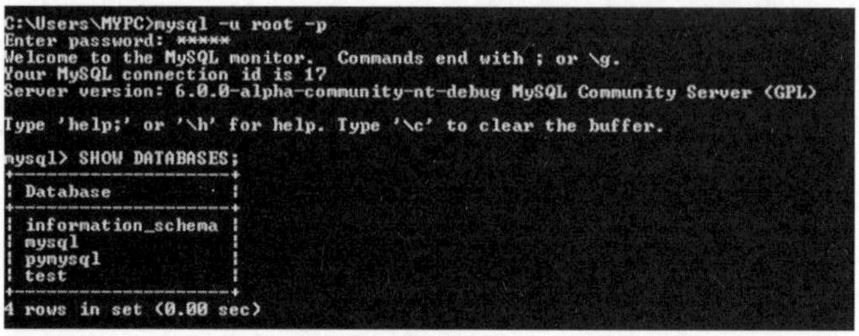

图 4.5

下面进行数据库授权。现在可创建用户，该用户对该数据库拥有所有权限。代码如下所示。

```
GRANT ALL PRIVILEGES ON pymysql.* TO 'pymysqladmin'@'localhost'
IDENTIFIED BY 'pymysql123';
```

请注意：
（1）本示例中对 pymysql 数据库具有所有权限的用户名为 pymysqladmin。
（2）密码为 pymysql123。
连接数据库需执行下面两个简单步骤。
步骤1：导入 mysql.connector。代码如下所示。

```
>>> import mysql.connector as msql
```

步骤2：打开数据库连接。
重点为使用 connect() 方法返回连接对象以连接 MySQL。
语法：

```
'your_connection_name = msql.connect(host = host_name,user = user_name,
passwd = database_password,charset ='utf8', database = database_name)
```

本示例使用 connect( )方法创建一个连接对象 pycon。代码如下所示。

```
>>>pycon = msql.connect(host = 'localhost',user ='pymysqladmin',pass-
wd ='pymysql123',charset ='utf8',database ='pymysql')
```

MySQL 6.0 中需提供参数 charset = 'utf8'。对于更高版本则无须提供。Python shell 中该段代码如图 4.6 所示。

```
>>> import mysql.connector as msql
>>> pycon = msql.connect(host = 'localhost',user='pymysqladmin',passwd='pymysql1
23',charset='utf8',database='pymysql')
```

图 4.6

注意：通过 is_connected( )函数检验 MySQL 数据库是否已建立连接。代码如下所示。

```
>>>pycon.is_connected()
True
>>>
```

步骤 3：创建游标对象。代码如下所示。

```
>>> mycursor = pycon.cursor()
```

创建游标非常重要，因为无论何时将搜索请求传递给数据库，都会返回一个必须存储的结果集。可以通过游标存储，如图 4.7 所示。

但游标存储存在限制。游标本质为只读，即不能使用游标更新数据库中的表，如图 4.8 所示。

图 4.7　　　　　　　　　　　图 4.8

按图 4.9 所示方法将游标赋值给变量。

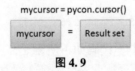

图 4.9

现在有了用于管理游标的变量 mycursor。在 Python 中游标通过 execute( )方法执行 SQL 查询。完成任务后，注意关闭游标对象，可使用 cursor.close( )方法完成。

步骤 4：创建表。

根据姓名簿创建包含以下 3 个字段的表。

（1）chapter_num 为自动增量与主键。

（2）chapter（用作章节名称）为文本字段且不为空。

(3) pages 为非空字段。

创建表的 MySQL 语句。代码如下所示。

```
CREATE TABLE book ( chapter_num INT UNSIGNED PRIMARY KEY AUTO_INCREMENT, chapter TEXT NOT NULL,pages INT NOTNULL )
```

在 Python 中，该语句借助游标执行。代码如下所示。

```
mycursor.execute("CREATE TABLE book ( chapter_num INT UNSIGNED PRIMARY KEY AUTO_INCREMENT, chapter TEXT NOT NULL,pages INT NOTNULL )")
```

使用命令提示符直接检查表是否已成功创建。
输入密码进行登录：

```
mysql -u pymysqladmin -p
Enter Password: pymysql123
```

现在需要对表进行操作，输入以下指令。

```
use pymysql;
```

检查表是否已创建，在命令提示符中输入以下指令。

```
show columns from book;
```

显示结果如图 4.10 所示。

图 4.10

使用 MySQL 查询命令在命令提示符中添加一条记录，并尝试使用 Python 检索该记录。

在命令提示符中输入以下指令。

```
insert into book values(1,'Introduction',12);
```

查看该记录是否已添加至表中。代码如下所示。

```
select * from book;
```

显示结果如图 4.11 所示。

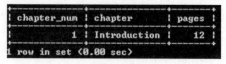

图 4.11

4.3.2 使用 Python 检索数据库记录

现在 book 表中存有一条记录。

在 Python LDLE 中，可以使用如下语法检索 book 表中的记录。

```
cursor_name.fetchall()
```

图 4.12 所示为使用 Python 连接 MySQL 服务器。

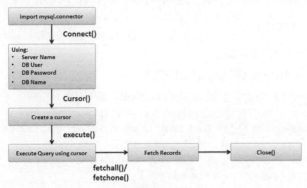

图 4.12

检索数据库记录的步骤如下。

步骤 1：导入 API 模块。代码如下所示。

```
>>> import mysql.connector as msql
```

步骤 2：连接数据库。代码如下所示。

```
>>> pycon = msql.connect(host = 'localhost',user ='pymysqladmin',passwd ='pymysql123',charset ='utf8',database ='pymysql')
```

步骤 3：通过连接对象创建游标。代码如下所示。

```
>>> mycursor = pycon.cursor()
```

步骤 4：使用 SQL 语句执行 MySQL 语句。代码如下所示。

```
>>> statement = 'select * from pymysql.book;'
>>>mycursor.execute(statement)
```

步骤 5：读取数据。代码如下所示。

```
>>>mycursor.fetchall()
```

使用 mycursor 检索记录，代码如下所示。

```
>>>import mysql.connector as msql
>>>pycon = msql.connect(host = 'localhost',user ='pymysqladmin',passwd ='pymysql123',charset ='utf8',database ='pymysql')
```

```
>>> mycursor = pycon.cursor()
>>> statement = 'select * from pymysql.book;'
>>> mycursor.execute(statement)
>>> mycursor.fetchall()
[(1,'Introduction',12)]
```

步骤6：关闭 connection( )。

尽管上文示例未执行此步骤，但其必不可少。代码如下所示。

```
>>> mycursor.close()
```

在下一个示例中执行该步骤。

直接将该记录与 MySQL 读取的记录匹配。

通过 Python Shell 插入记录。进入 Python Shell，输入以下指令。

```
>>> mycursor.execute("""INSERT INTO book(chapter, pages) VALUES("Basics", "15");""")
>>> pycon.commit()
>>> statement = 'select * from book;'
>>> mycursor.execute(statement)
>>> mycursor.fetchall()
[(1,'Introduction',12),(6,'Basics',15)]
```

由此可见，已添加第二条记录。可直接通过命令提示符再次查看，如图4.13所示。

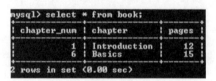

图 4.13

至此已广泛探讨了 MySQL 查询过程。下面介绍如何在 MySQL Bash 上进行基本的 MySQL 查询，以及在 Python 中进行相同的查询。

4.4 创建数据库

使用 Python 创建数据库，重点是使用 MySQL 创建数据库（在 4.4.1 小节中学习），4.4.2 小节将介绍如何使用 Python 创建 MySQL 数据库。

4.4.1 使用 MySQL 创建数据库

按照以下步骤使用 MySQL 创建 SQL 数据库。

步骤1：在命令提示符中输入以下指令。

```
mysql -u root -p
```

步骤 2：输入密码。

系统提示输入使用 MySQL 的密码。安装时已设置此密码，本示例中为 bpb12。图 4.14 所示为输入密码以连接 MySQL。

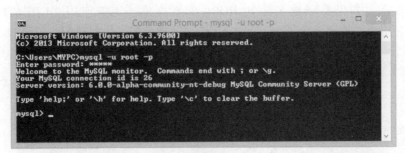

图 4.14

步骤 3：输入以下指令创建数据库。

mysql > CREATE DATABASE SCHOOL;

按图 4.15 所示指令查看刚创建的数据库是否在列表中。

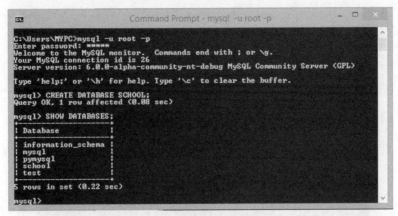

图 4.15

通过 Python 接口进行相同操作。为此，需删除数据库，再检查其是否已被删除。

DROP DATABASE SCHOOL;

输入 SHOW DATABASES 指令，确认其是否被删除。删除数据库 SCHOOL 后的数据库列表如图 4.16 所示。

图 4.16

4.4.2　使用 Python 创建数据库

本小节介绍如何使用 Python 创建与 4.4.1 小节中相同的数据库。现在读者已经熟悉所涉及的步骤。如果要创建名为 SCHOOL 的数据库，则需使用"CREATE database SCHOOL;"指令。使用 mycursor.execute( ) 将相同指令传递给数据库。代码如下所示。

```
import mysql.connector as msql
pycon = msql.connect(host = 'localhost',user ='root',passwd ='bpb12',
charset ='utf8')
mycursor = pycon.cursor()
statement = 'CREATE database SCHOOL;'
mycursor.execute(statement)
mycursor.execute("SHOW DATABASES;")
mycursor.fetchall()
mycursor.close()
```

Python Shell 中该段代码如图 4.17 所示。比较图 4.17 与图 4.18 中的输出。

```
>>> import mysql.connector as msql
>>> pycon = msql.connect(host = 'localhost',user='root',passwd='bpb12',charset='utf8')
>>> mycursor = pycon.cursor()
>>> statement = 'CREATE database SCHOOL;'
>>> mycursor.execute(statement)
>>> mycursor.execute("SHOW DATABASES;")
>>> mycursor.fetchall()
[('information_schema',), ('mysql',), ('pymysql',), ('school',), ('test',)]
```

图 4.17

MySQL 执行指令"SHOW DATABASES;"，查看输出结果，如图 4.18 所示。

图 4.18

使用 Python 以相同方式删除数据库，代码如下所示。

```
#删除数据库
statement = 'DROP DATABASE SCHOOL;'
mycursor.execute(statement)
mycursor.execute("SHOW DATABASES;")
mycursor.fetchall()
mycursor.close()
```

在 Python Shell 与 MySQL Bash 中查看数据库相关信息的方式,如图 4.19 所示。

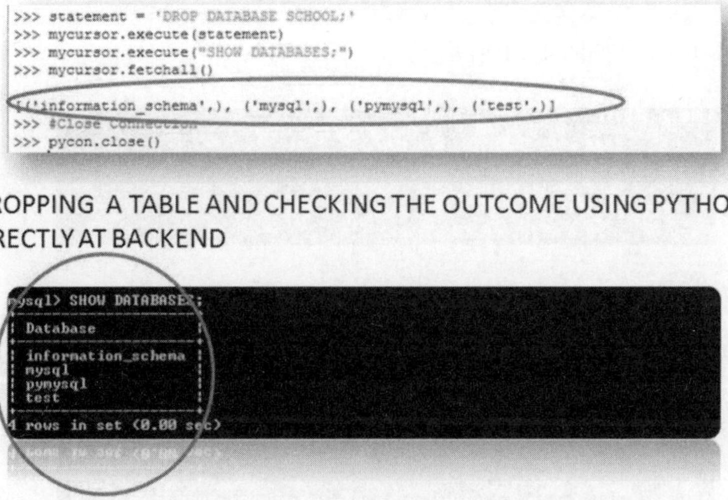

图 4.19

示例 4.1

使用 Python 创建表格 BOOK_LIBRARY。

注意:操作本章最初创建的 pymysql 数据库。根据姓名簿创建表,并在数据库中创建表 BOOK_LIBRARY。

配置 MySQL 后为 pymysql 数据库设置了权限。数据库用户名为 pymysqladmin,密码为 pymysql123,如图 4.20 所示。

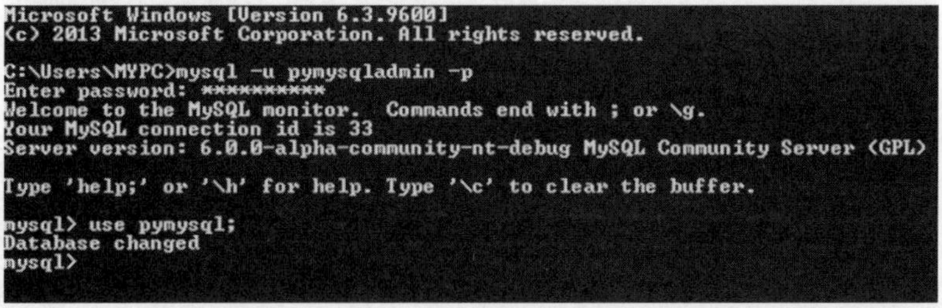

图 4.20

现在学习如何在 MySQL 上直接创建表 BOOK_LIBRARY。代码如下所示。

```
CREATE TABLE BOOK_LIBRARY(
    BOOK_ID SMALLINT NOT NULL AUTO_INCREMENT,
    BOOK_NAME CHAR(20) NOT NULL,
    AUTHOR_NAME CHAR(20) NOT NULL,
    PAGES SMALLINT NOT NULL,
    COST SMALLINT NOT NULL,
```

```
    PRIMARY KEY(BOOK_ID)
);
```

其在 MySQL Bash 中如图 4.21 所示。

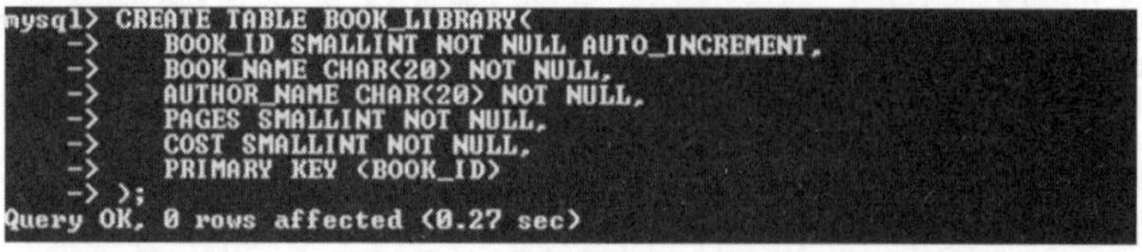

图 4.21

输入 SHOW COLUMNS FROM BOOK_LIBRARY 指令查看表格，如图 4.22 所示。

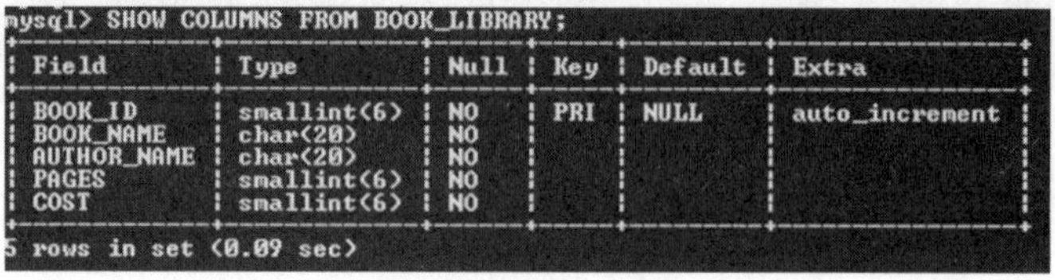

图 4.22

尝试使用 Python 得到相同结果。但在此之前，需要从数据库中删除表，如图 4.23 所示。

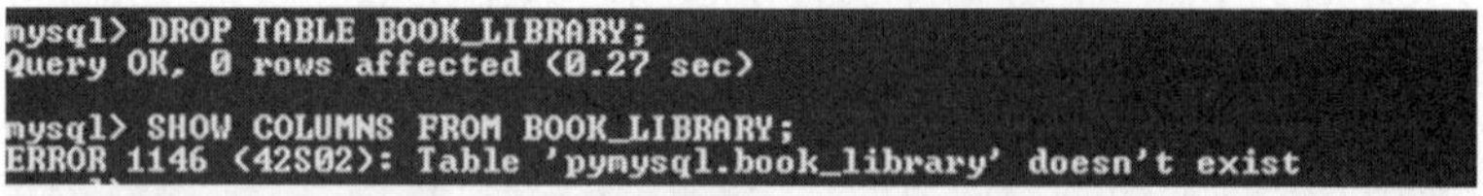

图 4.23

使用 Python 创建表的代码如下所示。

```
import mysql.connector as msql
pycon = msql.connect(host = 'localhost',user ='pymysqladmin',passwd ='py-
mysql123',charset ='utf8',database ='pymysql')
mycursor = pycon.cursor()
statement = """CREATE TABLE BOOK_LIBRARY(BOOK_ID SMALLINT UNSIGNED
PRIMARY KEY AUTO_INCREMENT, BOOK_NAME CHAR(20) NOT NULL,AUTHOR_NAME
CHAR(20) NOT NULL, PAGES SMALLINT NOT NULL,COST SMALLINT NOT NULL )"""
mycursor.execute(statement)
statement = 'SHOW COLUMNS FROM BOOK_LIBRARY;'
mycursor.execute(statement)
mycursor.fetchall()
    pycon.close()
```

Python Shell 与 MySQL Bash 中的输出如图 4.24 所示。

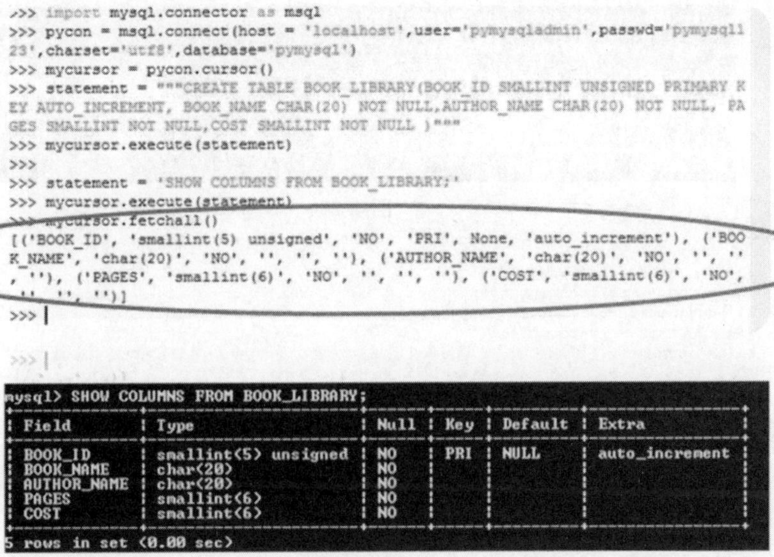

```
>>> import mysql.connector as msql
>>> pycon = msql.connect(host = 'localhost',user='pymysqladmin',passwd='pymysql1
23',charset='utf8',database='pymysql')
>>> mycursor = pycon.cursor()
>>> statement = """CREATE TABLE BOOK_LIBRARY(BOOK_ID SMALLINT UNSIGNED PRIMARY K
EY AUTO_INCREMENT, BOOK_NAME CHAR(20) NOT NULL,AUTHOR_NAME CHAR(20) NOT NULL, PA
GES SMALLINT NOT NULL,COST SMALLINT NOT NULL )"""
>>> mycursor.execute(statement)
>>>
>>> statement = 'SHOW COLUMNS FROM BOOK_LIBRARY;'
>>> mycursor.execute(statement)
>>> mycursor.fetchall()
[('BOOK_ID', 'smallint(5) unsigned', 'NO', 'PRI', None, 'auto_increment'), ('BOO
K_NAME', 'char(20)', 'NO', '', '', ''), ('AUTHOR_NAME', 'char(20)', 'NO', '', ''
, ''), ('PAGES', 'smallint(6)', 'NO', '', '', ''), ('COST', 'smallint(6)', 'NO',
 '', '', '')]
>>>
```

```
mysql> SHOW COLUMNS FROM BOOK_LIBRARY;
+-------------+----------------------+------+-----+---------+----------------+
| Field       | Type                 | Null | Key | Default | Extra          |
+-------------+----------------------+------+-----+---------+----------------+
| BOOK_ID     | smallint(5) unsigned | NO   | PRI | NULL    | auto_increment |
| BOOK_NAME   | char(20)             | NO   |     |         |                |
| AUTHOR_NAME | char(20)             | NO   |     |         |                |
| PAGES       | smallint(6)          | NO   |     |         |                |
| COST        | smallint(6)          | NO   |     |         |                |
+-------------+----------------------+------+-----+---------+----------------+
5 rows in set (0.00 sec)
```

图 4.24

下面学习如何在此表中插入和更新记录。

4.5　使用 Python 操作数据库

本节介绍如何使用 Python 进行以下操作。
（1）使用 INSERT 语句插入记录。
（2）读取记录。
（3）使用 WHERE 子句读取表中选定记录。
（4）使用 ORDER BY 子句对查询结果进行排序。
（5）使用 DELETE 指令删除记录。
（6）使用 UPDATE 指令更新记录。

4.5.1　使用 INSERT 语句插入记录

使用 INSERT 语句将记录插到表中。图 4.25 所示为使用 INSERT 语句直接在 MySQL 数据库中插入记录。

MySQL INSERT 语句如下所示。

```
INSERT INTO BOOK_LIBRARY(BOOK_NAME,AUTHOR_NAME,PAGES,COST) VALUES (' The Secret','Rhonda Byrne',198,3);
```

使用 Python 将记录插入数据库，如图 4.26 所示。
注意：使用 pycon.commit( )保存数据库中的更改。
查看 Python Shell 与 MySQL Bash 中的结果，如图 4.27 所示。

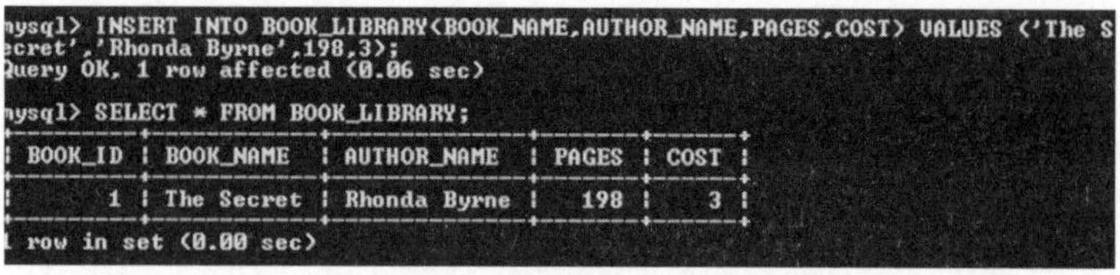

```
mysql> INSERT INTO BOOK_LIBRARY(BOOK_NAME,AUTHOR_NAME,PAGES,COST) VALUES ('The S
ecret','Rhonda Byrne',198,3);
Query OK, 1 row affected (0.06 sec)

mysql> SELECT * FROM BOOK_LIBRARY;
+---------+------------+-------------+-------+------+
| BOOK_ID | BOOK_NAME  | AUTHOR_NAME | PAGES | COST |
+---------+------------+-------------+-------+------+
|       1 | The Secret | Rhonda Byrne|   198 |    3 |
+---------+------------+-------------+-------+------+
1 row in set (0.00 sec)
```

图 4.25

```
>>> import mysql.connector as msql
>>> pycon = msql.connect(host = 'localhost',user='pymysqladmin',passwd='pymysql1
23',charset='utf8',database='pymysql')
>>> mycursor = pycon.cursor()
>>> statement = """INSERT INTO BOOK_LIBRARY(BOOK_NAME,AUTHOR_NAME,PAGES,COST) VA
LUES ('The Power','Rhonda Byrne',272,6);"""
>>> mycursor.execute(statement)
>>> pycon.commit()
```

图 4.26

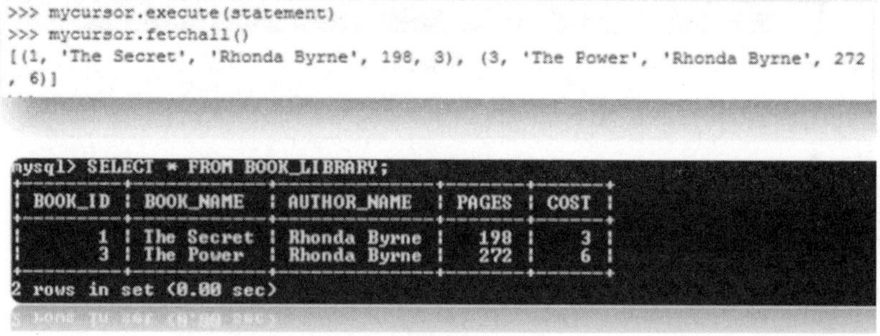

```
>>> mycursor.execute(statement)
>>> mycursor.fetchall()
[(1, 'The Secret', 'Rhonda Byrne', 198, 3), (3, 'The Power', 'Rhonda Byrne', 272
, 6)]
```

```
mysql> SELECT * FROM BOOK_LIBRARY;
+---------+------------+-------------+-------+------+
| BOOK_ID | BOOK_NAME  | AUTHOR_NAME | PAGES | COST |
+---------+------------+-------------+-------+------+
|       1 | The Secret | Rhonda Byrne|   198 |    3 |
|       3 | The Power  | Rhonda Byrne|   272 |    6 |
+---------+------------+-------------+-------+------+
2 rows in set (0.00 sec)
```

图 4.27

示例 4.2

在 BOOK_LIBRARY 表中动态插入记录。

代码:

```python
import mysql.connector as msql
data =[]
pycon = msql.connect(host = 'localhost',user ='pymysqladmin',passwd =
'pymysql123',charset ='utf8',database ='pymysql')
mycursor = pycon.cursor()
statement = """INSERT INTO BOOK_LIBRARY(BOOK_NAME,AUTHOR_NAME,PAGES,COST)
VALUES (%s,%s,%s,%s);"""
#读取书名称
val1 = input("Enter Book Name : ")

#将其附加至列表
data.append(val1)

#读取作者名
val2 = input("Enter Author Name : ")
```

```
#将其附加至列表
data.append(val2)

#读取页数并转换为 int
val3 = int(input("Enter Number of Pages: "))

#将其附加至列表
data.append(val3)

#读取成本值并转换为 int
val4 = int(input("Enter Cost in $ : "))

#将其附加至列表
data.append(val4)

#将列表转换为元组
data2 = tuple(data)

print(data2)
mycursor.execute(statement,data2)
pycon.commit()
pycon.close()
```

输出:

```
Enter Book Name : The Magic
Enter Author Name : Rhonda Byrne
Enter Number of Pages: 272
Enter Cost in $ : 6
('The Magic','Rhonda Byrne',272,6)
```

查看数据库中是否显示新记录,如图 4.28 所示。

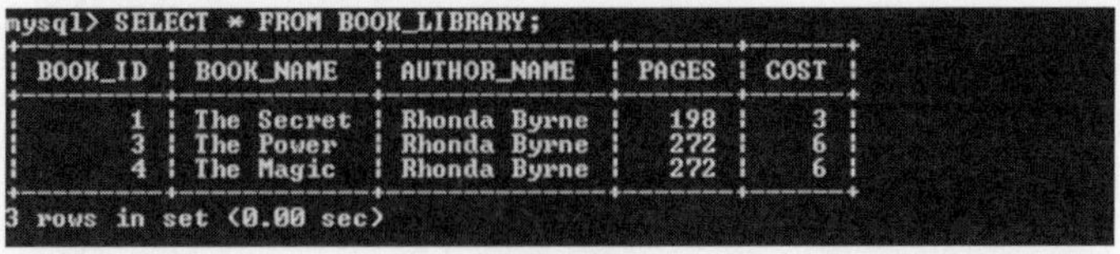

图 4.28

注意:下面的代码使用 format( )函数代替%格式。自主选择即可。
代码:

```
import mysql.connector as msql
pycon = msql.connect(host ='localhost',user ='pymysqladmin',passwd ='py-mysql123',charset ='utf8',database ='pymysql')
mycursor = pycon.cursor()
statement = """INSERT INTO BOOK_LIBRARY(BOOK_NAME,AUTHOR_NAME,PAGES,COST) VALUES ('{0}','{1}',{2},{3});"""
```

```
val1 = input("Enter Book Name : ")
val2 = input("Enter Author Name : ")
val3 = int(input("Enter Number of Pages: "))
val4 = int(input("Enter Cost in $ : "))
mycursor.execute(statement.format(val1,val2,val3,val4))
pycon.commit()
pycon.close()
```

输出：

```
Enter Book Name : The Hero
Enter Author Name : Rhonda Byrne
Enter Number of Pages: 240
Enter Cost in $: 6
```

结果如图 4.29 所示。

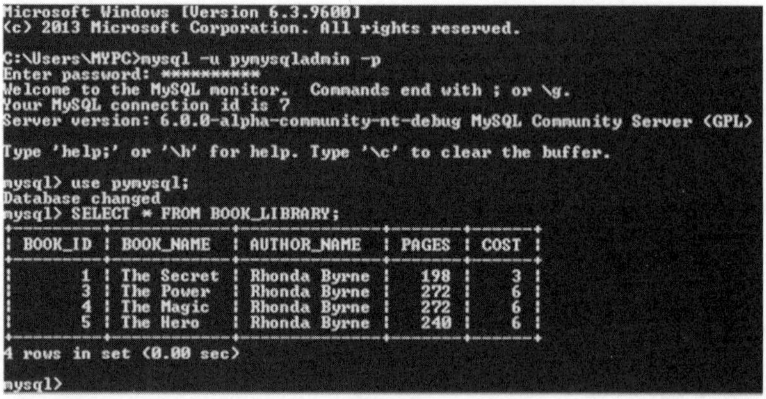

图 4.29

有时操作数据库需要将多个记录同时添加。为此，使用 executemany( )方法。通过以下实例学习其实现原理。假设需要在 BOOK_LIBRARY 表中添加 5 条记录，如表 4.1 所示。

表 4.1

BOOK_NAME	AUTHOR_NAME	PAGES	COST
I AM Magic	Maria Robins	63	13
One Truth, One Law	Erin Werley	87	6
Mind Magic	Merlin Starlight	265	19
Advanced Manifesting	Linda West	172	4
The Frequency	Linda West	180	4

步骤 1：将每条记录放入括号中。代码如下所示。

```
('I AM Magic','Maria Robins',63,13)
('One Truth, One Law','Erin Werley',87,6)
```

```
('Mind Magic','Merlin Starlight',265,19)
('Advanced Manifesting','Linda West',172,4)
('The Frequency','Linda West',180,4)
```

步骤2：创建包含所有记录的列表。代码如下所示。

```
values =[('I AM Magic','Maria Robins',63,13),('One Truth, One Law','Erin Werley',87,6),('Mind Magic','Merlin Starlight',265,19),('Advanced Manifesting','Linda West',172,4),('The Frequency','Linda West',180,4)]
```

步骤3：执行代码。

通过该列表进行添加。代码如下所示。

```
import mysql.connector as msql
pycon = msql.connect(host ='localhost',user ='pymysqladmin',passwd ='pymysql123',charset ='utf8',database ='pymysql')
mycursor = pycon.cursor()
statement = """INSERT INTO BOOK_LIBRARY(BOOK_NAME,AUTHOR_NAME,PAGES,COST) VALUES (%s,%s,%s,%s);"""
values =[('I AM Magic','Maria Robins',63,13),('One Truth, One Law','Erin Werley',87,6),('Mind Magic','Merlin Starlight',265,19),('Advanced Manifesting','Linda West',172,4),('The Frequency','Linda West',180,4)]
mycursor.executemany(statement,values)
pycon.commit()
pycon.close()
print("Records Inserted!!")
```

输出：

Records Inserted!!

4.5.2 读取记录

使用 Python 从表中读取数据表所有记录。

执行"SELECT * FROM BOOK_LIBRARY;"语句，查询显示 BOOK_LIBRARY 表中的所有记录（见图4.30）。下面使用 Python 读取数据表所有记录。

图 4.30

1. 使用 fetchall( ) 函数读取数据表所有记录

代码：

```
import mysql.connector as msql
pycon = msql.connect(host ='localhost',user ='pymysqladmin',passwd ='py-mysql123',charset ='utf8',database ='pymysql')
mycursor = pycon.cursor()
statement = """SELECT * FROM BOOK_LIBRARY;"""
mycursor.execute(statement)
result_set = mycursor.fetchall()
print("The Results are as follows: ")
for result in result_set:
    print(result)
pycon.close()
print("Done!!")
```

输出结果如图 4.31 所示。

```
The Results are as follows:
(1, 'The Secret', 'Rhonda Byrne', 198, 3)
(3, 'The Power', 'Rhonda Byrne', 272, 6)
(4, 'The Magic', 'Rhonda Byrne', 272, 6)
(5, 'The Hero', 'Rhonda Byrne', 240, 6)
(8, 'I AM Magic', 'Maria Robins', 63, 13)
(9, 'One Truth, One Law', 'Erin Werley', 87, 6)
(10, 'Mind Magic', 'Merlin Starlight', 265, 19)
(11, 'Advanced Manifesting', 'Linda West', 172, 4)
(12, 'The Frequency', 'Linda West', 180, 4)
Done!!
```

图 4.31

2. 使用 fetchone( ) 函数读取数据表单条记录

代码：

```
import mysql.connector as msql
pycon = msql.connect(host ='localhost',user ='pymysqladmin',passwd ='py-mysql123',charset ='utf8',database ='pymysql')
mycursor = pycon.cursor()
statement = """SELECT * FROM BOOK_LIBRARY;"""
mycursor.execute(statement)
result_set = mycursor.fetchone()
print(result_set)
pycon.close()
print("Done!!")
```

输出：

```
(1,'The Secret','Rhonda Byrne',198,3)
Done!!
```

示例 4.3

从表中检索选定列。

使用 SQL 直接从 MySQL 中检索选定记录,如图 4.32 所示。

图 4.32

使用 fetchall( ) 函数检索列表。

代码:

```
import mysql.connector as msql
pycon = msql.connect(host = 'localhost',user ='pymysqladmin',passwd ='py-
mysql123',charset ='utf8',database ='pymysql')
mycursor = pycon.cursor()
statement = """SELECT BOOK_NAME, AUTHOR_NAME FROM BOOK_LIBRARY;"""
mycursor.execute(statement)
result_set = mycursor.fetchall()
print("The Results are as follows: ")
for result in result_set:
    print(result)
pycon.close()
print("Done!!")
```

输出结果如图 4.33 所示。

```
The Results are as follows:
('The Secret', 'Rhonda Byrne')
('The Power', 'Rhonda Byrne')
('The Magic', 'Rhonda Byrne')
('The Hero', 'Rhonda Byrne')
('I AM Magic', 'Maria Robins')
('One Truth, One Law', 'Erin Werley')
('Mind Magic', 'Merlin Starlight')
('Advanced Manifesting', 'Linda West')
('The Frequency', 'Linda West')
Done!!
```

图 4.33

3. 使用 fetchone( ) 函数检索首行选定列

代码:

```
import mysql.connector as msql
pycon = msql.connect(host = 'localhost',user ='pymysqladmin',passwd ='py-
mysql123',charset ='utf8',database ='pymysql')
mycursor = pycon.cursor()
statement = """SELECT BOOK_NAME, AUTHOR_NAME FROM BOOK_LIBRARY;"""
```

```
mycursor.execute(statement)
result_set = mycursor.fetchone()
print(result_set)
pycon.close()
print("Done!!")
```

输出:

```
('The Secret','Rhonda Byrne')
Done!!
```

4.5.3 使用 WHERE 子句读取表中选定记录

WHERE 子句用于根据条件筛选记录。假设检索所有售价大于或等于 7 美元的图书信息。执行以下指令。

```
SELECT * FROM BOOK_LIBRARY WHERE COST >= 7;
```

售价大于或等于 7 美元的图书信息如图 4.34 所示。

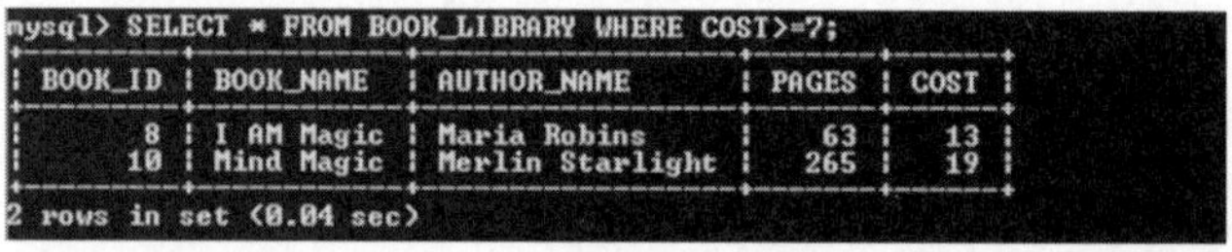

```
mysql> SELECT * FROM BOOK_LIBRARY WHERE COST>=7;
+---------+------------+-----------------+-------+------+
| BOOK_ID | BOOK_NAME  | AUTHOR_NAME     | PAGES | COST |
+---------+------------+-----------------+-------+------+
|       8 | I AM Magic | Maria Robins    |    63 |   13 |
|      10 | Mind Magic | Merlin Starlight|   265 |   19 |
+---------+------------+-----------------+-------+------+
2 rows in set (0.04 sec)
```

图 4.34

在 Python 中使用 WHERE 子句。

代码:

```
import mysql.connector as msql
pycon = msql.connect(host = 'localhost',user ='pymysqladmin',passwd ='py-mysql123',charset ='utf8',database ='pymysql')
mycursor = pycon.cursor()
statement = """SELECT * FROM BOOK_LIBRARY WHERE COST >=7;"""
mycursor.execute(statement)
result_set = mycursor.fetchall()
for result in result_set:
    print(result)
pycon.close()
print("Done!!")
```

为了显示从数据库中检索到的数据，使用 mycursor 从游标中读取所有记录，并赋给另一变量 result_set，然后遍历 result_set 以显示数据、执行其他操作或更新基础数据，如图 4.35 所示。

输出:

```
(8,'I AM Magic','Maria Robins', 63, 13)
(10,'Mind Magic','Merlin Starlight', 265, 19)
Done!!
```

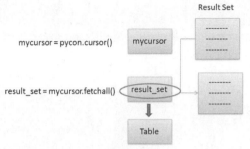

图 4.35

示例 4.4

从 BOOK_LIBRARY 表中选取所有作者名中包含 Rhon 的记录，如图 4.36 所示。

图 4.36

代码：

```
import mysql.connector as msql
pycon = msql.connect(host ='localhost',user ='pymysqladmin',passwd ='py-mysql123',charset ='utf8',database ='pymysql')
mycursor = pycon.cursor()
statement = """SELECT * FROM BOOK_LIBRARY WHERE AUTHOR_NAME LIKE '%Rhon%';"""
mycursor.execute(statement)
result_set = mycursor.fetchall()
for result in result_set:
    print(result)
pycon.close()
print("Done!!")
```

输出：

```
(1,'The Secret','Rhonda Byrne',198,3)
(3,'The Power','Rhonda Byrne',272,6)
(4,'The Magic','Rhonda Byrne',272,6)
(5,'The Hero','Rhonda Byrne',240,6)
Done!!
```

使用占位符%s 转义查询值。

代码：

```
import mysql.connector as msql
pycon = msql.connect(host ='localhost',user ='pymysqladmin',passwd ='py-mysql123',charset ='utf8',database ='pymysql')
```

```
mycursor = pycon.cursor()
statement = """SELECT * FROM BOOK_LIBRARY WHERE AUTHOR_NAME = %s;"""
author = ("Rhonda Byrne",)
mycursor.execute(statement,author)
result_set = mycursor.fetchall()
for result in result_set:
    print(result)
pycon.close()
print("Done!!")
```

输出：

```
(1,'The Secret','Rhonda Byrne',198,3)
(3,'The Power','Rhonda Byrne',272,6)
(4,'The Magic','Rhonda Byrne',272,6)
(5,'The Hero','Rhonda Byrne',240,6)
Done!!
```

4.5.4　使用 ORDER BY 子句对查询结果进行排序

代码：

```
import mysql.connector as msql
pycon = msql.connect(host ='localhost',user ='pymysqladmin',passwd ='py-mysql123',charset ='utf8',database ='pymysql')
mycursor = pycon.cursor()
statement = """SELECT * FROM BOOK_LIBRARY WHERE AUTHOR_NAME = %s;"""
author = ("Rhonda Byrne",)
mycursor.execute(statement,author)
result_set = mycursor.fetchall()
for result in result_set:
    print(result)
pycon.close()
print("Done!!")
```

输出：

```
(11,'Advanced Manifesting','Linda West',172,4)
(8,'I AM Magic','Maria Robins',63,13)
(10,'Mind Magic','Merlin Starlight',265,19)
(9,'One Truth, One Law','Erin Werley',87,6)
(12,'The Frequency','Linda West',180,4)
(5,'The Hero','Rhonda Byrne',240,6)
(4,'The Magic','Rhonda Byrne',272,6)
(3,'The Power','Rhonda Byrne',272,6)
(1,'The Secret','Rhonda Byrne',198,3)
Done!!
```

结果如图4.37所示。

图4.37

4.5.5 使用DELETE指令删除记录

目前读者已学习了如何在数据库中插入和检索记录。在本小节读者将学习如何从数据库中删除记录。

代码：

```
import mysql.connector as msql
pycon = msql.connect(host = 'localhost',user ='pymysqladmin',passwd ='py-mysql123',charset ='utf8',database ='pymysql')
mycursor = pycon.cursor()
statement = """DELETE FROM BOOK_LIBRARY WHERE BOOK_ID = 8"""
mycursor.execute(statement)
pycon.commit()
pycon.close()
print("Done!!")
```

输出结果如图4.38所示。

图4.38

4.5.6 使用UPDATE指令更新记录

代码：

```
import mysql.connector as msql
pycon = msql.connect(host = 'localhost',user ='pymysqladmin',passwd ='py-mysql123',charset ='utf8',database ='pymysql')
mycursor = pycon.cursor()
statement = """UPDATE BOOK_LIBRARY SET BOOK_ID = 2 WHERE BOOK_ID = 12;"""
mycursor.execute(statement)
pycon.commit()
```

```
pycon.close()
print("Done!!")
```

输出结果如图 4.39 所示。

图 4.39

小结

本章介绍了如何使用 Python 与数据库交互。可以使用 Python 执行任何类型的 SQL 查询。在第 9 章中读者将进一步了解如何使用 GUI 与数据库交互。因为有标准步骤，所以使用 Python 与数据库交互十分简单。一旦理解了实现步骤，就可以轻松编写任何类型的应用程序与数据库交互。

编程简答题

1. 编写 Python 代码，创建数据库 TEXTILE。

答：

```
>>> import mysql.connector as msql
>>>pycon = msql.connect(host = 'localhost',user ='root',pass-
wd = 'bpb12',charset ='utf8')
>>>mycursor = pycon.cursor()
>>> statement = 'CREATE database TEXTILE;'
>>>mycursor.execute(statement)
>>>pycon.commit()
>>>pycon.close()
```

在数据库中验证结果，如图 4.40 所示。

图 4.40

2. 给定用户 id 为 shopkeeper、密码为 shoptoday 的数据库 textile。编写 Python 代码连接 textile 数据库，如图 4.41 所示。

```
mysql> GRANT ALL PRIVILEGES ON TEXTILE.* TO 'shopkeeper'@'localhost' IDENTIFIED
BY 'shoptoday';
Query OK, 0 rows affected (0.34 sec)

mysql> quit
Bye

C:\Users\MYPC>mysql -u shopkeeper -p
Enter password: *********
Welcome to the MySQL monitor.  Commands end with ; or \g.
Your MySQL connection id is 36
Server version: 6.0.0-alpha-community-nt-debug MySQL Community Server (GPL)

Type 'help;' or '\h' for help. Type '\c' to clear the buffer.

mysql> use textile;
Database changed
mysql>
```

图 4.41

答：

```
import mysql.connector as msql
pycon = msql.connect(host = 'localhost',user =' shopkeeper',passwd = 'shoptoday',database ='textile')
```

3. 公司创建新数据库记录员工信息。员工信息如图 4.42 所示。新员工 Karan 加入公司担任项目经理。其员工编号为 7980，工资为 100 万元。将其信息以系统日期为入职日期插入表格。编写相关代码。

```
+-------+---------+--------------------+------------+---------+
| EmpNo | EmpName | Job                | Hiredate   | Sal     |
+-------+---------+--------------------+------------+---------+
| 7110  | Rai     | Director           | 2013-08-11 | 2500000 |
| 7123  | Rishab  | Software Developer | 2020-01-01 |  500000 |
| 7128  | Lee     | Software Developer | 2018-03-01 |  500000 |
| 7801  | Ray     | Temp               | 2020-07-01 |   60000 |
+-------+---------+--------------------+------------+---------+
```

图 4.42

答：

```
import mysql.connector as msql
from datetime import datetime
pycon = msql.connect(host = 'localhost',user ='root',passwd ='password',charset ='utf8',database = 'database_name')
mycursor = pycon.cursor()
now = datetime.now()
formatted_date = now.strftime('%Y-%m-%d')
statement = "INSERT INTO emp VALUES(7980,'Karan','Project Manager','{}',1000000);".format(formatted_date)
```

```
mycursor.execute(statement)
pycon.commit()
pycon.close()
```

在数据库中验证结果，如图 4.43 所示。

```
+-------+---------+-------------------+------------+---------+
| EmpNo | EmpName | Job               | Hiredate   | Sal     |
+-------+---------+-------------------+------------+---------+
| 7110  | Rai     | Director          | 2013-08-11 | 2500000 |
| 7123  | Rishab  | Software Developer| 2020-01-01 | 500000  |
| 7128  | Lee     | Software Developer| 2018-03-01 | 500000  |
| 7801  | Ray     | Temp              | 2020-07-01 | 60000   |
| 7980  | Karan   | Project Manager   | 2020-07-15 | 1000000 |
+-------+---------+-------------------+------------+---------+
rows in set (0.00 sec)
```

图 4.43

4. 公司决定所有员工减薪 10 000 元。根据上题，操作表 emp 实现数据更新。

答：

```
import mysql.connector as msql
pycon = msql.connect(host = 'localhost',user ='root',passwd = 
'password',charset ='utf8',database = 'database_name')
mycursor = pycon.cursor()
statement = 'UPDATE emp SET Sal = Sal - 10000;'
mycursor.execute(statement)
pycon.commit()
pycon.close()
```

输出结果如图 4.44 所示。

```
mysql> Select * from emp;
+-------+---------+-------------------+------------+---------+
| EmpNo | EmpName | Job               | Hiredate   | Sal     |
+-------+---------+-------------------+------------+---------+
| 7110  | Rai     | Director          | 2013-08-11 | 2490000 |
| 7123  | Rishab  | Software Developer| 2020-01-01 | 490000  |
| 7128  | Lee     | Software Developer| 2018-03-01 | 490000  |
| 7801  | Ray     | Temp              | 2020-07-01 | 50000   |
| 7980  | Karan   | Project Manager   | 2020-07-15 | 990000  |
+-------+---------+-------------------+------------+---------+
```

图 4.44

5. Modern fashion 等主流网店销售各式时尚男女服装，如图 4.45 所示。受新冠肺炎疫情影响，女装打 5 折，男装打 4 折。使用 Python 接口将此变化反映到 MySQL 数据库。

```
mysql> select * from modernFashions;
+----------+-----------+---------+------+
| item_num | item_name | for_who | cost |
+----------+-----------+---------+------+
| 101      | maxi      | women   | 2500 |
| 102      | jackets   | women   | 4500 |
| 103      | jackets   | men     | 6000 |
| 104      | patiala   | women   | 900  |
| 105      | pyjamas   | men     | 999  |
| 106      | scarfs    | women   | 3900 |
| 107      | skirts    | women   | 5900 |
| 109      | Jeans     | men     | 3999 |
+----------+-----------+---------+------+
8 rows in set (0.00 sec)
```

图 4.45

答：

```
import mysql.connector as msql
    pycon = msql.connect(host = 'localhost',user ='root',passwd =
'password',charset ='utf8',database = 'database_name')
    mycursor = pycon.cursor()
    statement = "UPDATE modernFashions SET cost = cost - 0.5 * cost where
for_who = '{}';".format('women')
    mycursor.execute(statement)
    pycon.commit()
    pycon.close()
```

输出结果如图 4.46 所示。

图 4.46

6. 表 emp 如图 4.47 所示。

图 4.47

2015 年以前入职的员工加薪 50 000 元。

答：

```
import mysql.connector as msql
    pycon = msql.connect(host = 'localhost',user ='root',passwd =
'password',charset ='utf8',database = 'database_name')
    mycursor = pycon.cursor()
    statement = "UPDATE emp SET sal = sal + 50000 where hiredate <=
'2015-01-01';".format('2015-01-01')
    mycursor.execute(statement)
    pycon.commit()
    pycon.close()
```

输出结果如图 4.48 所示。

```
: EmpNo : EmpName : Job               : Hiredate   : Sal     :
:  7110 : Rai     : Director          : 2013-08-11 : 2590000 :
:  7123 : Rishab  : Software Developer: 2020-01-01 :  540000 :
:  7128 : Lee     : Software Developer: 2018-03-01 :  540000 :
:  7801 : Ray     : Temp              : 2020-07-01 :  100000 :
:  7980 : Karan   : Project Manager   : 2020-07-15 : 1040000 :
```

图 4.48

7. 解释以下数据库对象方法。

a. close( )

b. cursor( )

c. rollback( )

答：

a. close( )方法用于关闭游标。

b. cursor( )方法返回一个数据库游标对象用于执行 SQL 查询。

c. rollback( )方法用于将待处理事务回滚到先前状态。

第 5 章

Python线程

> **引言** 进程是软件开发人员使用的常用术语。系统中任何运行程序均为进程。线程是进程或子进程的一部分,即分配处理器时间的基本单元。线程存在于进程中,并与进程共享同一内存空间。进程可有多个线程。本章介绍 Python 中线程的用法。

知识结构

- 进程和线程
- 创建线程
- 使用 Lock 和 RLock 实现线程同步
- Lock 的用法
- 死锁
  - 使用 Locked( )方法检查资源锁定状态
  - RLock
- 信号量
- 使用事件对象同步线程
- 条件类
- 后台线程和非后台线程

目标

完成本章的学习后,读者将掌握以下知识。
(1) 区分进程和线程;
(2) 创建线程;
(3) 线程同步信息;
(4) 了解 Lock、RLock、死锁、符号;
(5) 理解后台线程和非后台线程之间的区别。

> 至此读者已经了解了 Python 语言在软件开发领域如此流行的诸多特性。下面回顾部分 Python 的重要特性。
> （1）具有清晰可读的语法；
> （2）有广泛的标准库；
> （3）易于学习；
> （4）支持快速开发且易于调试；
> （5）提供基于异常的错误处理；
> （6）有完备的文档；
> （7）有良好的用户群体。

5.1 进程和线程

进程在计算机编程中很常见。本节介绍什么是进程，及其与线程有何不同。

1. 进程

进程是由操作系统创建、正在执行程序的实例。双击系统程序或应用程序（如 Word 文档、浏览器等）可启动进程。每个应用程序启动自身进程，包含自身数据栈、空间地址和其他辅助数据，该数据允许应用程序跟踪自身执行。同一个 Python 程序的多个进程可以同时执行代码。

操作系统负责管理进程执行，调度进程，分配系统计算资源。

2. 线程

线程是轻量级活动控制流，可与其他线程并行激活，也可在同一进程中与其他线程并发执行。进程中可运行多个线程，其中每个线程均为独立程序，与其他线程一起运行。

线程可以将程序本身分割成两个或多个并发运行的任务。每个线程独立执行任务，与同一进程中的其他线程并行。线程与同一进程中其他线程共享寻址空间和数据结构（即线程读写同一变量），是程序中的一组语句，其执行不依赖其余代码。

线程化允许执行多个线程，如任务、函数调用等，同时，线程化不是并行编程。事实上，当任务执行包含某种等待时（例如，与 Web 服务器上其他机器交互时），需要等待才能建立连接。在这段等待时间中可执行其他代码。在同一 Python 程序中，两个线程不可同时执行代码。

注意：与进程不同，线程共享进程的资源，多个线程可以共享同一地址空间和其他资源，进程之间的信息共享比线程之间的信息共享慢。

线程的使用难点是多个线程使用同一块数据。下面介绍如何使用线程。

5.2 创建线程

线程是标准库，无须另外下载。在使用线程之前，需要导入 threading 模块。
使用线程的步骤如下。

步骤 1：导入 threading 模块。
这是使用线程代码编写的第一条语句，如下所示。

```
import threading
```

步骤 2：定义目标函数。
定义线程调用的目标函数。代码如下所示。

步骤 3：实例化线程。
在实例化线程时，必须提供步骤 2 中定义的函数作为目标函数。代码如下所示。

```
thread_name = threading.Thread(target = function_name,args = (i,))
```

最简单的线程使用方法是通过目标函数将其实例化，Python 线程模块提供了 Thread( ) 方法。其语法格式如下所示。

```
class threading.Thread( group = name, target = None, name = None, args = (), kwargs = {} )
```

（1）group：为将来实现而保留，取值为 None。
（2）target：线程启动时执行的函数名。
（3）name：允许为线程命名。在默认情况下，Python 为线程提供新名称，格式为 Thread – N。
（4）args：向目标函数传递值的元组。
（5）kwargs：用作目标函数关键字参数的字典，为可选关键字参数目录。

步骤 4：启动线程。
如果要使线程工作，需调用 start( )方法。start( )方法允许线程开始工作。代码如下所示。

```
thread_name.start()
```

每个线程对象最多只能调用一次该方法。以下是关于 start( )方法的主要用法。
（1）用于在 Windows 和 Linux 中创建新线程。
（2）启动线程并使其通过传递参数列表（实例化线程时供给元组 args）调用目标

函数。

（3）元组 args 将参数转发至目标函数。如果函数不需要元组，则使用空元组。

步骤 5：调用 join( ) 方法。代码如下所示。

```
thread_name.join()
```

join( ) 方法用于阻塞当前线程，直至其调用对象阻塞为止。join( ) 方法从主线程中调用。一旦调用，join( ) 方法将防止主线程比其调用线程先退出。

注：将在示例 5.2 中说明 join( ) 方法的重要性。

示例 5.1

编写程序显示线程名称。

步骤 1：导入 threading 模块。代码如下所示。

```
import threading
```

步骤 2：定义目标函数。代码如下所示。

```
def thread_count(count):
    print("I am the thread number ",count,".")
```

调用函数 thread_count( ) 显示线程数量。

步骤 3：实例化线程。代码如下所示。

```
for i in range(1,6):
    t = threading.Thread(target = thread_count,args = (i,))
```

实例化 5 个线程，count(i) 作为参数传递给目标函数 thread_count( )。

步骤 4：启动线程。代码如下所示。

```
t.start()
```

步骤 5：调用 join( ) 方法。代码如下所示。

```
t.join()
```

代码：

```
import threading

def thread_count(count):
    print("I am the thread number ",count,".")

for i in range(1,6):
    t = threading.Thread(target = thread_count,args = (i,))
    t.start()
    t.join()
```

代码说明如图 5.1 所示。

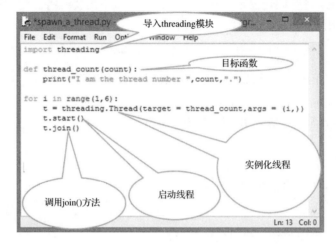

图 5.1

输出：

```
I am the thread number 1 .
I am the thread number 2 .
I am the thread number 3 .
I am the thread number 4 .
I am the thread number 5 .
>>>
```

示例 5.2

说明 join( )方法的功能。

答：本示例在示例 5.1 的代码末尾添加一行代码，如图 5.2 所示。

图 5.2

输出结果如图 5.3 所示。

```
I am the thread number  1 .
I am the thread number  2 .
I am the thread number  3 .
I am the thread number  4 .
I am the thread number  5 .
Main thread exiting now
```

图 5.3

由此可见，主线程等待所有线程结束后，打印其退出语句"Main thread exiting now"。

注释 t. join( )语句并执行代码，如图 5.4 所示。

```
File  Edit  Format  Run  Options  Window  Help
import threading

def thread_count(count):
    print("I am the thread number ",count,".")

for i in range(1,6):
    t = threading.Thread(target = thread_count,args = (i,))
    t.start()
#    t.join()
print("Main thread exiting now")
                                                        Ln: 17  Col: 0
```

图 5.4

输出结果如图 5.5 所示。

```
I am the thread number Main thread exiting nowI am the thread number I am the th
read number I am the thread number I am the thread number
    1
>>> 2345     .....
```

图 5.5

主线程先于其他线程退出。

示例 5.3

编写代码确定当前线程。

如果是非显式命名线程，则以格式 Thread - N 为其指定唯一名称。其中，N 是线程编号。因此，线程名称为 Thread - 1、Thread - 2、Thread - 3 等。读者可通过本示例学习如何确定当前线程。

步骤 1：导入 threading 模块。代码如下所示。

```
import threading
```

步骤 2：定义目标函数。代码如下所示。

```
def thread_count(count):
    print("I am the thread number ",count,".")
    print("My name is ",threading.currentThread().getName())
```

currentThread( )函数用于返回与调用者控制线程对应的当前线程对象。当前线程调用 getName( )函数返回线程名称。

步骤3：实例化线程。代码如下所示。

```
for i in range(1,6):
    t = threading.Thread(target = thread_count,args = (i,))
```

步骤4：启动线程。代码如下所示。

```
t.start()
```

步骤5：调用 join( )方法。代码如下所示。

```
t.join()
```

代码：

```
import threading

def thread_count(count):
    print("I am the thread number ",count,".")
    print("My name is ",threading.currentThread().getName())
for i in range(1,6):
    t = threading.Thread(target = thread_count,args = (i,))
    t.start()
    t.join()
```

输出：

```
I am the thread number  1 .
My name is  Thread-1
I am the thread number  2 .
My name is  Thread-2
I am the thread number  3 .
My name is  Thread-3
I am the thread number  4 .
My name is  Thread-4
I am the thread number  5 .
My name is  Thread-5
```

示例5.4

假设陪客人去俱乐部。俱乐部接待客人并生成用于晚间住宿的 guest_id。每接待1位客人需等待5秒，提供1个 guest_id 需等待10秒。为此情况编写代码。

步骤1：导入 time 模块。代码如下所示。

```
import time
```

步骤2：创建函数记录客人姓名。

创建函数接收客人姓名。代码如下所示。

```python
names = [ ]
def getnames():
    guest_name = input("Enter a guest name : ")
    names.append(guest_name)
    confirmation = input("Any more guests? (y/n) : ")
if(confirmation == 'y'):
    getnames()
```

步骤3：创建接待函数。代码如下所示。

```python
def welcome_guests(names):
    for name in names:
        print("Welcome to our club",name)
        time.sleep(5)
```

步骤4：为每位客人分配 guest_id。代码如下所示。

```python
def get_id(names):
    count =1
    for name in names:
        print("The guest_id of ",name,"is :", count)
        count =   count +1
        time.sleep(10)
```

代码：

```python
import time
names = [ ]

def getnames():
    guest_name = input("Enter a guest name : ")
    names.append(guest_name)
    confirmation = input("Any more guests? (y/n) : ")
    if(confirmation == 'y'):
        getnames()

def welcome_guests(names):
    for name in names:
        print("Welcome to our club",name)
        time.sleep(5)

def get_id(names):
```

```
            count =1
            for name in names:
                print("The guest_id of ",name,"is :", count)
                count =   count +1
                time.sleep(10)

#执行
getnames()
welcome_guests(names)
get_id(names)
```

输出：

```
Enter a guest name : Raj
Any more guests? (y/n) : y
Enter a guest name : Ted
Any more guests? (y/n) : y
Enter a guest name : Reggie
Any more guests? (y/n) : n
Welcome to our club Raj
Welcome to our club Ted
Welcome to our club Reggie
The guest_idof   Raj is : 1
The guest_idof   Ted is : 2
The guest_idof   Reggie is : 3
```

现在间隔 5 秒生成一条欢迎信息，间隔 10 秒生成一条身份信息。

示例 5.5

使用 threading 模块完成示例 5.4 中的任务。

代码：

```
import time
import threading
names = [ ]

def getnames():
    guest_name = input("Enter a guest name : ")
    names.append(guest_name)
    confirmation = input("Any more guests? (y/n) : ")
    if(confirmation == 'y'):
        getnames()

def welcome_guests(names):
    for name in names:
        print("Welcome to our club",name)
        time.sleep(5)
```

```
def get_id(names):
    count = 1
    for name in names:
        print("The guest_id of ",name,"is :", count)
        count =  count + 1
        time.sleep(10)

getnames()
t = time.time()

#创建线程
t1 = threading.Thread(target = welcome_guests, args = (names,))
t2 = threading.Thread(target = get_id, args = (names,))

#启动线程
t1.start()
time.sleep(2)
t2.start()

#阻塞进程
t1.join()
t2.join()
```

输出：

```
Enter a guest name : Raj
Any more guests? (y/n) : y
Entera guest name : Ted
Any more guests? (y/n) : y
Enter a guest name : Reggie
Any more guests? (y/n) : n
Welcome to our club Raj
The guest_id of  Raj is : 1
Welcome to our club Ted
Welcome to our club Reggie
The guest_id of  Ted is : 2
The guest_id of  Reggie is : 3
All Done enjoy
```

下面学习如何使用 threading 模块创建新线程。

使用 threading 模块创建新线程，需要创建 thread 类的子类。期间只能使用以下两种方法更改：__init__( )方法和 run( )方法。

（1）重写 run 方法。

步骤1：导入 time 和 threading 模块。代码如下所示。

```
import time
import threading
```

步骤 2：定义新线程类的子类并重写 run( ) 方法。
定义子类。代码如下所示。

```
class learning_subclassing(threading.Thread):
```

这一步重写原来的 run( ) 方法。threading.py 中 run( ) 方法的原始代码如下所示。

```
try:
    if self._target:
        self._target(*self._args, **self._kwargs)
finally:
    # 如果线程运行的函数具有指向线程的成员的参数,则避免重复循环
    del self._target, self._args, self._kwargs
```

只在原始代码基础上添加 print 语句。因此，run( ) 方法的目前代码如下所示。

```
def run(self):
    print('\n thread {} has started'.format(self.getName()),"\n")#新增代码
    try:
        if self._target:
            self._target(*self._args, **self._kwargs)
    finally:
        #如果线程运行的函数具有指向线程的成员的参数,则避免重复循环
        del self._target, self._args, self._kwargs
```

run( ) 方法成为线程起点。

步骤 3：定义目标函数。

该函数有两个参数：sec 和 name。其中，name 是线程名称；sec 是以秒为单位的时间。在该线程中，name 线程休眠秒数为 sec。代码如下所示。

```
def nap(sec,name):
    print("\n Hi I am a thread and my name is {} . I am here only to take a nap of {} seconds".format(name,sec),"\n")
    time.sleep(sec)
    print("{} seconds are over {} is up!!".format(sec,name),"\n")
```

步骤 4：创建新线程。
调用 Thread 类的构造函数。代码如下所示。

```
t1 = learning_subclassing(target = nap, name = "King-thread",args = (10,"King-thread",))
t2 = learning_subclassing(target = nap, name = "Queen-thread",args = (6,"Queen-thread"))
t3 = learning_subclassing(target = nap, name = "Page-thread",args = (7,"Page-thread"))
```

步骤 5：调用 start( ) 函数启动线程执行。代码如下所示。

```
t1.start()
t2.start()
t3.start()
```

步骤 6：调用 join( ) 方法。代码如下所示。

```
t1.join()
t2.join()
t3.join()
```

代码：

```
import time
import threading

class learning_subclassing(threading.Thread):
    def run(self):
        print('\n thread {} has started'.format(self.getName()),
        "\n")
        try:
            if self._target:
                self._target(*self._args, **self._kwargs)
        finally:
            del self._target, self._args, self._kwargs

def nap(sec,name):
    print("\n Hi I am a thread and my name is {}. I am here only to
    take a nap of {} seconds".format(name,sec),"\n")
    time.sleep(sec)
    print("{} seconds are over {} is up!!".format(sec,name),"\n")

t1 = learning_subclassing(target = nap, name = "King-thread",args =
(10,"King-thread",))
t2 = learning_subclassing(target = nap, name = "Queen-thread",args =
(6,"Queen-thread"))
t3 = learning_subclassing(target = nap, name = "Page-thread",args =
(7,"Page-thread"))
t1.start()
t2.start()
t3.start()
t1.join()
t2.join()
t3.join()
```

输出：

```
thread King-thread has started
thread Queen-thread has started
```

```
thread Page - thread has started

    Hi I am a thread and my name is King - thread. I am here only to take a nap of
10 seconds
    Hi I am a thread and my name is Queen - thread. I am here only to take a nap of
6 seconds
    Hi I am a thread and my name is Page - thread. I am here only to take a nap of
7 seconds

    6 seconds are over Queen - thread is up!!

    7 seconds are over Page - thread is up!!

    10 seconds are over King - thread is up!!

>>>
```

（2）重写__init__( )和run( )方法。

下面通过重写threading模块的__init__( )和run( )方法实现相同的示例。

步骤1：导入语句。代码如下所示。

```
import time
import threading
```

步骤2：定义learn_thread类继承threading. Thread类。代码如下所示。

```
class Learn_thread(threading.Thread):
```

步骤3：重写构造函数。

构造函数必须调用基类的__init__( )函数。代码如下所示。

```
class Learn_thread(threading.Thread):
    def __init__(self,name,sec):
        #调用基类的__init__()函数
        threading.Thread.__init__(self)
        self.name = name
        self.sec = sec
```

步骤4：定义nap( )函数。代码如下所示。

```
def nap(sec,name):
    print("\n Hi I am a thread and my name is {}. I am here only to
    take a nap of {} seconds".format(name,sec),"\n")
    time.sleep(sec)
    print("{} seconds are over {} is up!!".format(sec,name),"\n")
```

步骤5：重写run( )方法。

由于已创建新__init__()函数，所以失去目标。run()方法将调用nap()函数。因此，在这里实现逻辑。代码如下所示。

```
#重写run()方法
def run(self):
    print('\n thread {} has started'.format(self.getName())," \n")
    nap(self.sec,self.name)
```

步骤6：创建线程实例并启动线程。代码如下所示。

```
t1 = Learn_thread("King-thread",10)
t2 = Learn_thread("Queen-thread",6)
t3 = Learn_thread("Page-thread",7)
t1.start()
t2.start()
t3.start()
t1.join()
t2.join()
t3.join()
```

代码：

```
import time
import threading


class Learn_thread(threading.Thread):
    def __init__(self,name,sec):
        #调用基类的__init__函数
        threading.Thread.__init__(self)
        self.name = name
        self.sec = sec

#重写run()方法
    def run(self):
        print('\n thread {} has started'.format(self.getName())," \n")
        nap(self.sec,self.name)

def nap(sec,name):
    print("\n Hi I am a thread and my name is {}. I am here only to
    take a nap of {} seconds".format(name,sec),"\n")
    time.sleep(sec)
    print("{} seconds are over {} is up!!".format(sec,name),"\n")

t1 = Learn_thread("King-thread",10)
t2 = Learn_thread("Queen-thread",6)
t3 = Learn_thread("Page-thread",7)
t1.start()
t2.start()
```

```
t3.start()
t1.join()
t2.join()
t3.join()
```

输出：

```
thread King-thread has started
thread Queen-thread has started
thread Page-thread has started

    Hi I am a thread and my name is King-thread. I am here only to take a nap of
10 seconds
    Hi I am a thread and my name is Queen-thread. I am here only to take a nap of
6 seconds
    Hi I am a thread and my name is Page-thread. I am here only to take a nap of
7 seconds

    6 seconds are over Queen-thread is up!!

    7 seconds are over Page-thread is up!!

    10 seconds are over King-thread is up!!

>>>
```

5.3 使用 Lock 和 RLock 实现线程同步

线程同步机制可以防止共享相同资源的两个或多个并发线程同一时间访问程序临界区。

图 5.6 所示为使用共享资源的 3 个线程，即 Thread-1、Thread-2 和 Thread-3。如果这 3 个线程每次只访问一个共享资源，则一切正常；否则当多个线程试图同时访问共享资源时，便会出现问题。此情况会导致竞争条件。

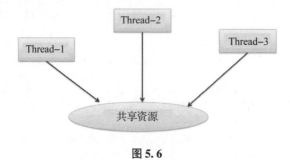

图 5.6

第 5 章　Python 线程

竞争条件是指两个或多个线程同时访问共享资源并试图更改数据的情况。在这种情况下，代码结果将不可预测。假如两个或更多线程同时访问同一全局变量并更改其值，这时无法确定哪个线程更改了变量值。

示例 5.6

解释说明竞争条件。

代码：

```
import threading

x = 0
range_val = int(input("Enter the range : "))

def thread_sync():
    global x
    for i in range(range_val):
        x += 1

t1 = threading.Thread(target = thread_sync)
t2 = threading.Thread(target = thread_sync)
t1.start()
t2.start()
t1.join()
t2.join()

print(x)
```

代码详解如图 5.7 所示。

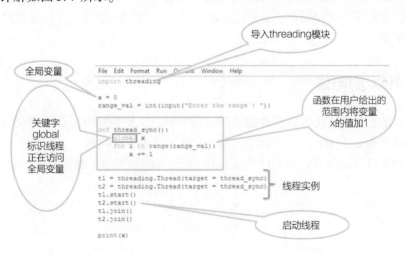

图 5.7

代码提示用户输入一个数，并将该值赋给变量 range_val。当调用线程时，将 range = range_val 中的 x 值加 1。

输出：
实例 1：对于 range_val = 10

线程 t1 执行 thread_sync( ) 函数，将 x 值加 10。线程 t2 也将 x 值加 10，因此显示输出为 20。输出正确。

```
Enter the range : 10
20
>>>
```

实例 2：对于 range_val = 1000

线程 t1 执行 thread_sync( ) 函数，将 x 值增加 1000。线程 t2 也将 x 值增加 1000，因此显示输出为 2000。输出正确。

```
Enter the range : 1000
2000
>>>
```

实例 3：对于 range = 100000

线程 t1 执行 thread_sync( ) 函数，并将 x 值递增 100000。线程 t2 还将 x 增加 100000，期望输出为 200000。但是，结果显示输出为 137540。输出不正确。

```
Enter the range : 100000
137540
>>>
```

在前两种情况下结果与预期一致，可能因为 range 值非常小。但是，当提供 100000 范围时，大值范围会导致线程同步。显然，这两个线程试图同时改变 x 值，这就是结果和预期不同的原因。

5.4　Lock 的用法

上一节介绍了线程同步以及可能出现的意外输出。使用同步工具处理竞争条件非常重要。使用 Python 中 threading 模块的 Lock 类也可以解决该问题。

Lock 对象分别通过 acquire( ) 和 release( ) 方法获取和释放，如图 5.8 所示。

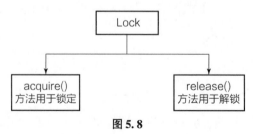

图 5.8

线程获取的 Lock 对象本质上可为阻塞或非阻塞。acquire()方法可以接收 True 和 False 为参数，默认值为 True。如果未指定参数，则 acquire()方法调用为阻塞。这也意味着 Lock 对象本质上取决于使用 True 还是 False 值调用 acquire()方法。使用 True 值（默认值）调用时，线程执行阻塞，直到解锁。如果获取值为 False，则线程执行非阻塞，直到将其设置为 True。

获得的 Lock 对象通过 release()方法释放。如果多个线程正在等待解锁，此方法只允许一个线程继续进行。如果在 Lock 对象已解锁时调用 release()方法，将导致运行时错误。

示例 5.7

使用与示例 5.6 相同的代码演示 Lock 的用法。

步骤 1：导入语句。代码如下所示。

```
import threading
```

步骤 2：声明全局变量。代码如下所示。

```
x = 0
range_val = int(input("Enter the range : "))
```

步骤 3：创建 Lock 类实例。代码如下所示。

```
lock = threading.Lock()
```

步骤 4：定义目标函数。

目标函数 thread_sync()有一个参数，即 Lock 对象。

注意：lock.acquire()在增加 x 值之前被调用。调用 acquire()方法时没有值，因此默认值为 True。一旦获得 Lock 对象的线程函数结束执行，就通过调用 release()方法释放 Lock，允许下一个线程处理该函数。函数代码与示例 5.6 中的函数代码略有不同，两者区别如下。

（1）acquire()方法在 x 值改变前被调用，这样每次只有一个线程访问全局变量 x。

（2）x 值改变后调用 release()方法解锁。

代码如下所示。

```
def thread_sync(lock):
    global x
    for i in range(range_val):
        lock.acquire()
        x += 1
        lock.release()
```

步骤 5：创建线程。代码如下所示。

```
t1 = threading.Thread(target = thread_sync, args = (lock,))
t2 = threading.Thread(target = thread_sync, args = (lock,))
```

以 target_sync( )作为目标函数创建线程，并将 Lock 类实例作为要传递给目标函数的参数。

步骤 6：调用 start( )方法。代码如下所示。

```
t1.start()
t2.start()
```

步骤 7：调用 join( )方法。代码如下所示。

```
t1.join()
t2.join()
```

代码：

```
import threading
x = 0
range_val = int(input("Enter the range : "))

def thread_sync(lock):
    global x
    for i in range(range_val):
        lock.acquire()

        x + = 1
        lock.release()

lock = threading.Lock()
t1 = threading.Thread(target = thread_sync, args = (lock,))
t2 = threading.Thread(target = thread_sync, args = (lock,))
t1.start()
t2.start()
t1.join()
t2.join()
print(x)
```

输出：

```
Enter the range : 1000000
2000000
>>>
```

如果尝试释放未锁定的 Lock 会怎么样，只需注释语句 lock. acquire( )，然后执行代码，观察输出结果。

5.5 死锁

到目前为止，读者已经学习了创建线程和线程的用法（使用线程有时会导致竞争条件问题），然后学习了如何使用锁避免竞争条件。在本节中读者将了解锁定机制可能导致的死锁情况。

假设两个线程（图5.9）Thread-1 和 Thread-2 有两个共享资源可供使用——共享 Resource1 和共享 Resource2。如果 Thread-1 访问共享 Resource1，而 Thread-2 访问共享 Resource2，两个线程都能获取锁并访问其资源。现在，假定 Thread-1 访问共享 Resource2，而 Thread-2 访问共享 Resource1，然后释放其锁。由于资源锁定，因此两个线程都被阻止访问其资源并继续等待。此情况称为死锁。

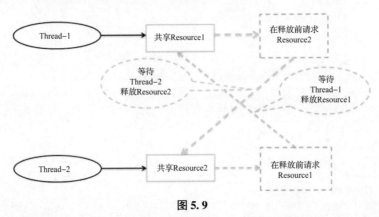

图5.9

以下代码演示死锁情况。两个函数 thread_sync1() 和 thread_sync2() 通过两个全局变量 x 和 y 执行不同计算。线程 t1 调用函数 thread_sync1()，获取锁1 并使用 x；线程 t2 调用函数 thread_sync2()，获取锁2 并使用 y。目前为止一切正常，但现在 t1 访问 y，并且 t2 访问 x，两个变量都被锁定，这便为死锁条件。

代码：

```
import threading
x = 0
y = 0

def thread_sync1(number):
    global x
    global y
    print("thread_sync1 acquiring lock1 \n")
    lock1.acquire()
    x = number*2
    print("thread_sync1 acquiring lock2 \n")
    lock2.acquire()
    z = x+y
    print("z =",z)
    print("thread_sync1 releasing lock2 \n")
    lock2.release()
    print("thread_sync1 releasing lock1 \n")
```

```
        lock1.release()
        print("thread_sync1 done \n")

def thread_sync2(number):
    global x
    global y
    print("thread_sync2 acquiring lock2 \n")

    lock2.acquire()
    y = number * 4
    print("thread_sync2 acquiring lock1 \n")
    lock1.acquire()
    v = y - x
    print("v = ",v)
    print("thread_sync2 releasing lock1 \n")
    lock1.release()
    print("thread_sync2 releasing lock2 \n")
    lock2.release()
    print("thread_sync2 done \n")

lock1 = threading.Lock()
lock2 = threading.Lock()
t1 = threading.Thread(target = thread_sync1, args =(2,))
t2 = threading.Thread(target = thread_sync2, args =(3,))
t1.start()
t2.start()
t1.join()
t2.join()
print("Final x = ",x)
print("Final y = ",y)
```

代码说明如图5.10所示。

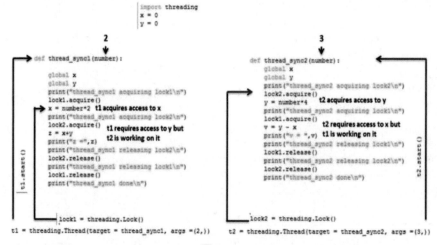

图5.10

输出：

```
thread_sync1 acquiring lock1
thread_sync2 acquiring lock2

thread_sync1 acquiring lock2
thread_sync2 acquiring lock1
```

5.5.1　使用 locked( )方法检查资源锁定状态

使用 locked( )方法检查锁定状态。此方法代码行已用粗体标记，以便识别。

代码：

```
import threading
x = 0
y = 0

def thread_sync1(number):
    global x
    global y
    print("thread_sync1 acquiring lock1\n")
    lock1.acquire()
    x = number * 2
    print("thread_sync1 acquiring lock2\n")
    if lock2.locked():
        print("in thread_sync1, the next resource is locked
        waiting\n")
    else:
        print("We have got the next resource")
        lock2.acquire()
        z = x + y
        print("z = ",z)
        print("thread_sync1 releasing lock2\n")
        lock2.release()
        print("thread_sync1 releasing lock1\n")
        lock1.release()
        print("thread_sync1 done\n")

def thread_sync2(number):
    global x
    global y
    print("thread_sync2 acquiring lock2\n")
    lock2.acquire()
    y = number * 4
    print("thread_sync2 acquiring lock1\n")
    if lock1.locked():
```

```
                print("in thread_sync2, the next resource is locked
                waiting\n")
            else:
                print("We have got the next resource")
                lock1.acquire()
                v = y - x
                print("v = ",v)
                print("thread_sync2 releasing lock1 \n")
                lock1.release()
                print("thread_sync2 releasing lock2 \n")
                lock2.release()
                print("thread_sync2 done \n")

lock1 = threading.Lock()
lock2 = threading.Lock()
t1 = threading.Thread(target = thread_sync1, args =(2,))
t2 = threading.Thread(target = thread_sync2, args =(3,))

t1.start()
t2.start()
t1.join()
t2.join()
```

输出：

```
thread_sync1 acquiring lock1
thread_sync2 acquiring lock2

thread_sync1 acquiring lock2
thread_sync2 acquiring lock1

in thread_sync1, the next resource is locked waiting
in thread_sync2, the next resource is locked waiting
```

Lock 可以为解决编程中某些问题提供解决方法，但也伴随某些问题。死锁就是典型示例。Lock 会引入不必要开销并影响代码的可伸缩性和可读性。锁定机制使代码难以调试。这就是使用替代方法同步访问共享内存的重要原因。

5.5.2 RLock 方法

本小节介绍重入锁（RLock）。顾名思义，同一线程多次获取 RLock。RLock 也有 acquire()和 release()方法，但与 Lock 有点不同，RLock 可以多次获取并且必须释放相同次数才能解锁。RLock 和 Lock 的另一个主要区别是，RLock 只能由获取它的线程释放，而 Lock 无法识别当前锁定的线程。

RLock 防止线程意外阻塞访问共享资源。RLock 中的共享资源可以被多个线程重复访问，RLock 是简单锁机制的稍高版本。在简单锁情况下，同一线程不能两次获取同一锁，而 RLock 对象不能被同一线程重新获取。

使用 RLock 解决死锁条件。

代码：

```
import threading
x = 0
y = 0

def thread_sync1(number):
    global x
    global y
    print("thread_sync1 acquiring lock1\n")
    lock.acquire()
    x = number*2
    print("thread_sync1 acquiring lock2\n")
    lock.acquire()
    z = x+y
    print("z =",z)
    print("thread_sync1 releasing lock2\n")
    lock.release()
    print("thread_sync1 releasing lock1\n")
    lock.release()
    print("thread_sync1 done\n")

def thread_sync2(number):
    global x
    global y
    print("thread_sync2 acquiring lock2\n")
    lock.acquire()
    y = number*4
    print("thread_sync2 acquiring lock1\n")
    lock.acquire()
    v = y - x
    print("v = ",v)
    print("thread_sync2 releasing lock1\n")
    lock.release()
    print("thread_sync2 releasing lock2\n")
    lock.release()
    print("thread_sync2 done\n")

lock = threading.RLock()
t1 = threading.Thread(target = thread_sync1,args =(2,))
t2 = threading.Thread(target = thread_sync2,args =(3,))

t1.start()
t2.start()
t1.join()
```

```
t2.join()
print("Final  x  =  ",x)
print("Final  y  =  ",y)
```

输出：

```
thread_sync1 acquiring lock
thread_sync2 acquiring lock
thread_sync1 acquiring lock

z = 4
thread_sync1 releasing lock
thread_sync1 releasing lock
thread_sync1 done
thread_sync2 acquiring lock

v = 8
thread_sync2 releasing lock
thread_sync2 releasing lock
thread_sync2 done

Final x =  4
Final y =  12
>>>
```

5.6　信号量

信号量是操作系统管理的抽象数据类型，为线程提供同步访问有限数量资源。信号量是内部变量，反映对其关联资源的并发访问次数。信号量的值始终大于 0 且小于或等于可用资源总数。另外，信号量包含 acquire( ) 和 release( ) 方法。

每次线程获取信号量同步资源时，信号量的值递减。同样，线程释放信号量同步资源时，信号量的值增加。信号量这个概念由荷兰计算机科学家 Edsger Dijkstra 提出。

根据信号量的初始化方式，多个线程可以同时访问同一代码。

使用信号量编写简单代码的步骤如下。

步骤1：导入语句。代码如下所示。

```
import threading
```

步骤2：初始化一个信号量。

对资源访问限制为 3 个线程。代码如下所示。

```
t = threading.Semaphore(3)
```

步骤3：定义函数。代码如下所示。

```
def count():
    t.acquire()
```

```
    print("Start")
    for i in range(1, 6):
        print(i)
    t.release()
```

这里定义简单函数 count( ) 来打印数字 1~5。

步骤 4：创建线程。

创建 6 个线程，它们的目标函数都是在步骤 3 中创建的 count( )。代码如下所示。

```
thread1 = threading.Thread(target = count)
thread2 = threading.Thread(target = count)
thread3 = threading.Thread(target = count)
thread4 = threading.Thread(target = count)
thread5 = threading.Thread(target = count)
thread6 = threading.Thread(target = count)
```

步骤 5：调用 start( ) 和 join( ) 方法。代码如下所示。

```
thread1.start()
thread2.start()
thread3.start()
thread4.start()
thread5.start()
thread6.start()

thread1.join()
thread2.join()
thread3.join()
thread4.join()
thread5.join()
thread6.join()
```

代码：

```
import threading

t = threading.Semaphore(3)
def count():
    t.acquire()
    print("Start")
    for i in range(1, 6):
        print(i)
    t.release()
thread1 = threading.Thread(target = count)
thread2 = threading.Thread(target = count)
```

```
thread3 = threading.Thread(target=count)
thread4 = threading.Thread(target=count)
thread5 = threading.Thread(target=count)
thread6 = threading.Thread(target=count)

thread1.start()
thread2.start()
thread3.start()
thread4.start()
thread5.start()
thread6.start()
thread1.join()
thread2.join()
thread3.join()
thread4.join()
thread5.join()
thread6.join()
```

信号量已将最大并发线程数限制为 3,因此输出分为 2 组。前 3 个线程同时执行函数,后 3 个线程稍后执行。

输出:

```
StartStartStart
>>>
111
222
333
444
555
StartStartStart
111
222
333
444
555
```

检查哪些线程同时执行。

代码:

```
import threading

t = threading.Semaphore(3)
def count():
    t.acquire()
    print("Starting ",threading.currentThread().getName())
```

```
    for i in range(1, 6):
        print(i)
    t.release()

thread1 = threading.Thread(target = count)
thread2 = threading.Thread(target = count)
thread3 = threading.Thread(target = count)
thread4 = threading.Thread(target = count)
thread5 = threading.Thread(target = count)
thread6 = threading.Thread(target = count)

thread1.start()
thread2.start()
thread3.start()
thread4.start()
thread5.start()
thread6.start()
thread1.join()
thread2.join()
thread3.join()
thread4.join()
thread5.join()
thread6.join()
```

输出：

```
Starting
>>> Starting  Starting   Thread-1Thread-2Thread-3

111
222
333
444
555
Starting StartingStarting    Thread-4Thread-5Thread-6
111
222
333
444
555
```

假设信号量允许7个并发线程，而不是3个如下所示。

```
t = threading.Semaphore(7)
```

在这种情况下，6个线程能够同时访问代码，输出结果如下所示。

```
StartStartStartStart
StartStartStartStart
111111
222222
333333
444444
555555
```

5.7 使用事件对象同步线程

本节介绍如何使用 Event 类对象进行线程通信，它有非常简单的机制，即其中一个线程发出事件信号，另一个线程等待信号。

事件对象使用称为事件标志的内部标志。使用 set( )方法将其设置为 True，同样，使用 clear( )方法将其设置为 False。Event 类的 wait( )方法可以阻止线程，直到事件标志设置为 True。

事件对象按以下方式初始化：

```
e = threading.Event()
```

在默认情况下，事件标志在初始化时设置为 False。

使用 isSet( )方法检查事件标志。代码如下所示。

```
>>> import threading
>>>evt = threading.Event()
>>>evt.isSet()
False
>>>
```

使用 set( )方法将事件标志设置为 True。代码如下所示。

```
>>> import threading
>>>evt = threading.Event()
>>>evt.isSet()
False
>>>evt.set()
>>>evt.isSet()
True
>>>
```

使用 clear( )方法将事件标志重置为 False。

```
>>> import threading
>>>evt = threading.Event()
```

```
>>> evt.isSet()
False
>>> evt.set()
>>> evt.isSet()
True
>>> evt.clear()
>>> evt.isSet()
False
>>>
```

wait()方法让线程等待事件。wait()方法只能在该事件上隐含,该事件的事件标志设置为 False。只要事件标志为 False,线程就等待。如果事件标志设置为 True,则无法阻止该线程。可以设置 wait()方法的 Timeout。代码如下所示。

```
wait([Timeout])
```

1. 事件标志设置为 False 的事件对象的代码
下面的代码演示当事件标志设置为 False 时事件对象的工作方式。

代码:

```
import threading
import time

#函数 task()以 event 和 sec 为参数
def task(event, sec):
    print("Started thread but waiting for event...")
    #使线程等待设置了超时的事件
    internal_set = event.wait(sec)
    if internal_set:
        print("Event set")
    else:
        print("Time out,event not set")

#初始化事件对象
e = threading.Event()

#启动线程,目标函数为 task,名称为 Event-Blocking-Thread
#传递一个事件对象,时间为 5 秒

t1 = threading.Thread(name='Event-Blocking-Thread', target=task,
args=(e,5))
t1.start()

#让主线程睡眠 3 秒
time.sleep(3)
```

输出:

```
Started thread but waiting for event...
>>> Time out,event not set
```

2. 事件标志设置为 True 的事件对象的代码

下面的代码演示当事件标志设置为 True 时事件对象的工作方式。

代码:

```
import threading
import time

def task(event, sec):
    print("Started thread but waiting for event...")
    #使线程等待设置了超时的事件
    internal_set = event.wait(sec)
    print("waiting")
    if internal_set:
        print("Event set")
    else:
        print("Time out,event not set")

#初始化事件对象
e = threading.Event()
print("Event is set.")
#启动线程
t1 = threading.Thread(name='Event-Blocking-Thread', target=task,
args=(e,10))
t1.start()
e.set()
```

输出:

```
Event is set.
Started thread but waiting for event...
>>>
waiting
Event set
```

既然已经理解事件对象的工作原理,调用 join() 方法后,如果事件标志设置为 True,则代码的输出结果如下所示。

```
import threading
import time

def task(event, sec):
    print("Started thread but waiting for event...")
```

```python
# 使线程等待设置了超时的事件
    internal_set = event.wait(sec)
    print("waiting")
    if internal_set:
        print("Event set")
    else:
        print("Time out,event not set")

# 初始化事件对象
e = threading.Event()
print('Event Object Created')
# 启动线程
t1 = threading.Thread(name ='Event-Blocking-Thread', target =task,
args =(e,10))
t1.start()
t1.join()
e.set()
```

join()方法可以暂停所有内容,直到task()函数执行完毕,调用join()方法后将事件标志设置为True。因此,当task()函数执行时,e未设置为True,输出结果如下所示。

```
Event is set.
Started thread but waiting for event...
waiting
Time out,event not set
```

因此,在调用join()方法前,将事件标志设置为True非常重要。

5.8 条件类

定义条件类是显著提高两个线程之间通信速度的方法。条件类的对象称为条件变量。

(1)条件变量使线程等待,直到出现条件。
(2)条件变量类似于事件对象的更高级版本。
(3)条件变量不能在进程之间共享。
(4)总是与某个锁关联。
(5)锁是条件变量的一部分,无须单独处理。

重要方法:
(1)notify(n-1):唤醒等待条件的线程,还提供要唤醒的线程编号。
(2)notify_all():唤醒所有等待条件的线程。

（3）wait([timeout = none])：等待通知或发生超时，如果调用notify()或notify_all()方法，则此方法将终止。如果调用此方法时线程尚未获取锁，则出现运行时错误。如果发生超时，则此方法返回False，否则将返回True。

代码：

```
import threading
import time
lst =[]

def buy():
    c.acquire()
    for i in range (1,6):
        item = input("Enter the item bought : ")
        lst.append(item)
    print("Notifying t2")
    c.notify()
    c.release()

def billing():
    c.acquire()
    c.wait(timeout = 0)
    c.release()
  print("you are billed for:")
    for item in lst:
        print(item)
c = threading.Condition()
t1 = threading.Thread(target = buy)
t2 = threading.Thread(target = billing)

t1.start()
time.sleep(5)
t2.start()
t1.join()
t2.join()
```

输出：

```
Enter the item bought : Rice
Enter the item bought : Wheat
Enter the item bought : Grain
Enter the item bought : Pen
Enter the item bought : Pencil
Notifying t2
you are billed for:
Rice
Wheat
Grain
Pen
Pencil
```

5.9 后台线程和非后台线程

线程有两种类型：后台线程和非后台线程。

后台线程不会阻止主线程退出，并继续在后台运行。相反，非后台线程阻止主线程退出，直到所有线程都退出。

后台线程为非后台线程提供支持。到目前为止，均只使用非后台线程。所有后台线程在非后台线程终止后都会终止。且必须都被明确终止。

线程状态只能在启动后设置。

有 3 种方法可以创建后台线程。

（1）第一种方法。

```
t1 = threading.Thread(target = method_name)
t1.setDaemon(True)
```

（2）第二种方法。

```
t = threading.Thread(target = method_name)
t.daemon = True
```

（3）第三种方法。

```
t = threading.Thread(target = method_name, daemon = True)
```

1. 设置后台线程并检查是否为 True

在下面示例中，使用 setDaemon()方法将线程设置为后台线程。线程启动之前，状态必须设置为 True。

（1）setDaemon(True)：设置线程为后台线程。

（2）setDaemon(False)：设置线程为非后台线程。

（3）setDaemon() 方法必须在 start()方法之前调用。如果在调用 start()方法之后调用，将生成运行时错误。

（4）借助 isDaemon()方法检查线程是否是后台程序。如果是后台线程，则返回 True，否则返回 False。

2. 检查后台线程的状态是否为 True

代码：

```
import threading
import time

def print_if_daemon():
    print("inside target function")
    if t1.isDaemon():
```

```python
                    print(threading.currentThread().name," is a daemon!!")
t1 = threading.Thread(target = print_if_daemon)
#检查 t1 是否是后台线程
print("Present Status of thread t1 is: ",t1.isDaemon())

#设置 t1 为后台线程
t1.setDaemon(True)

#调用 start()方法
t1.start()

time.sleep(5)
print("Main Exiting")
t1.join()
```

输出：

```
Present Status of thread t1 is:  False
inside target function
Thread-1is a daemon!!
Main Exiting
```

最后一个后台线程在所有非后台线程终止后自动终止。

3. 后台线程未设置为 True 的情形

代码：

```python
import threading
import time


def work_for_t1():

    print('Starting of thread :', threading.current_thread().name)
    for i in range(0,11):
        print("Printing Number : ",i," \n")
        time.sleep(2)
    print('Finishing of thread :', threading.current_thread().name,
    "\n")

t1 = threading.Thread(target =work_for_t1, name ='Thread1')
t1.start()
print("thread is Daemon : ", t1.daemon)
t1.join()
time.sleep(5)
print("Finishing thread: ", threading.current_thread().name," \n")
```

输出：

```
thread is Daemon : Starting of thread :   False Thread1
Printing Number :  0
Printing Number :  1
Printing Number :  2
Printing Number :  3
Printing Number :  4
Printing Number :  5
Printing Number :  6
Printing Number :  7
Printing Number :  8
Printing Number :  9
Printing Number :  10
Finishing of thread : Thread1
Finishing thread:  MainThread
```

由于 t1 是非后台线程，所以阻止主线程退出，直到其任务完成。

4. 后台线程设置为 True 的情形

主线程在打印 2 后结束，但 Thread1 继续。

通过设置后台线程为 True，即主线程终止后，后台线程继续执行。

代码：

```
import threading
import time

def work_for_t1():

    print('Starting of thread :', threading.current_thread().name)
    for i in range(0,11):
        print("Printing Number : ",i," \n")
        time.sleep(2)
    print('Finishing of thread :', threading.current_thread().name,
    " \n")


t1 = threading.Thread(target =work_for_t1, name ='Thread1')
t1.setDaemon(True)
t1.start()
time.sleep(5)
print("Finishing thread: ",threading.current_thread().name," \n")
```

输出中，主线程在 t1（后台线程）之前终止。

输出：

```
Starting of thread : Thread1
Printing Number :  0
```

```
Printing Number : 1
Printing Number : 2
Finishing thread: MainThread
Printing Number : 3
Printing Number : 4
Printing Number : 5
Printing Number : 6
Printing Number : 7
Printing Number : 8
Printing Number : 9
Printing Number : 10
Finishing of thread : Thread1
```

5. threading 模块的主要方法

threading 模块的主要方法如下。

（1） threading. activeCount( )/threading. active_count( )：返回活动线程数。

（2） threading. currentThread( )/threading. current_thread( )：返回当前线程。

（3） threading. enumerate( )：返回所有当前活动线程的列表，包括后台线程、当前线程创建的虚拟线程对象和主线程。已终止的线程不在列表中。

（4） threading. main_thread( )：返回启动 Python 解释器的主线程。

threading 模块中主要方法的实现如下。

代码：

```
import threading

def thread_count(count):
    print("I am the thread number ",count,".")
    print("My name is ",threading.currentThread().getName())
    print("active count",threading.activeCount())
    l1 = threading.enumerate()
    for i in l1:
        print(i)

for i in range(1,3):
    t = threading.Thread(target = thread_count,args = (i,))
    t.start()
    t.join()
```

输出：

```
I am the thread number  1 .
My name is  Thread-1
active count 3
<_MainThread(MainThread, started 7188)>
<Thread(SockThread, started daemon 7884)>
<Thread(Thread-1, started 10208)>
```

```
I am the thread number  2 .
My name is   Thread-2
active count 3
<_MainThread(MainThread, started 7188) >
< Thread(SockThread, started daemon 7884) >
< Thread(Thread-2, started 5648) >
>>>
```

6. 主线程的性质

Python 主线程是非后台线程。通过在 Python Shell 中输入以下语句进行检查。

```
>>>import threading
>>>threading.current_thread().getName()
'MainThread'
>>>threading.current_thread().isDaemon()
False
>>>
```

小结

　　本章介绍了如何使用 threading 模块处理 Python 线程。可通过简化设计实现 Python 线程并通过允许某些代码部分并发运行加快执行速度。

简答题

1. 以下横线上的内容应为什么？
   (1) 操作系统创建_____运行程序。
   (2) 一个进程可以有多个_____。
   (3) 线程_____进程内。
   (4) 线程共享_____空间。
   (5) _____模块提供最基本的线程同步机制。
   (6) 无论何时，锁均由_____线程或非线程持有。
   (7) 一旦锁被_____，其他线程就阻止访问共享资源，直到锁_____。
   (8) 可通过两个步骤① _____和② _____获得锁。
   (9) 如果要解锁，请调用_____方法。
   (10) 使用_____方法唤醒等待条件的所有线程。
   (11) 用 threading 模块创建新线程，必须创建_____类的子类。
   答：
   (1) 进程；

（2）线程；
（3）包含在；
（4）内存；
（5）threading；
（6）单；
（7）获得，释放；
（8）lock = Lock( )，lock.acquire( )；
（9）release( )；
（10）notifyAll( )；
（11）Thread。

2. 为什么线程称为轻量级进程？

答：

（1）线程是进程的一部分，因此具有许多共同特征。
（2）相比它所属的进程，其执行不繁重。

3. 什么是 GIL？

答： GIL 代表全局解释器锁。每次允许 1 个线程控制 Python 解释器，也就是每次只执行 1 个线程，采用锁或互斥防止多个线程同时执行来保护 Python 对象。GIL 非常重要，Python 解释器线程并不安全。为支持多线程编程，当前线程访问 Python 对象前保持全局锁非常重要。对于多线程解释器，定期释放并重新获取锁（在默认情况下，每 10 个字节指令代码后，此值可更改）。

4. 比较进程和线程的区别。

答：

进程与线程的区别如表 5.1 所示。

表 5.1 进程与线程的区别

进程	线程
独立执行顺序	独立执行顺序
在单独内存空间中运行	在共享内存空间中运行
具有 CPU 功能	具有操作环境功能
进程是正在执行的程序	线程是操作系统分配处理时间的基本单位
流程繁重，因为需要更多上下文切换时间	流程轻量，因此切换时间更短

5. 说明以下叙述的正误。

（1）线程是彼此并行运行的独立程序。
（2）线程在进程中运行。
（3）2 个进程比 2 个线程更容易彼此共享数据。

(4) 进程比线程需要更多的内存开销。

(5) 线程让程序更快，即使正在占用 100% CPU。

答：

(1) 正确；

(2) 正确；

(3) 错误；

(4) 正确；

(5) 错误。

6. 为什么编程中需要线程？

答：需要线程加快进程。线程用于在进程后台执行快速任务。

7. threading 模块的用途。

答：threading 模块是创建、控制和管理线程所必需的。

8. threading 模块包含实现线程的 Thread 类。简要说明 Thread 类的主要方法。

答：Thread 类提供的主要方法如下。

(1) run( )：线程入口点。

(2) start( )：启动线程。start( ) 方法可以调用 run( ) 方法，该方法是线程的入口点。

(3) join( )：阻止调用线程或主线程，直到调用其 join( ) 方法的线程终止为止。

(4) isAlive( )：检查线程是否结束执行。

(5) getName( )：返回线程名称。

(6) setName( )：设置线程名称。

9. Python 中的 threading 模块包括哪些不同类型对象？

答：Python 中的 threading 模块包括以下类型对象。

(1) 线程：表示执行的单个线程。

(2) Lock：锁对象。Lock 对象分别使用 acquire( ) 和 release( ) 方法获取和释放锁。

(3) RLock：重入锁对象。顾名思义，RLock 通过同一线程多次获取。RLock 也有 acquire( ) 和 release( ) 方法，但与 Lock 略有不同。RLock 可以多次获取锁并且必须释放相同次数才能解锁。RLock 和 Lock 的另一个主要区别是 RLock 只能由获取它的线程释放，而 Lock 不同，它无法识别当前持有锁的线程。

(4) 条件：定义条件类是显著提高两个线程通信速度的方法。条件类的对象称为条件变量。

①条件变量使线程等待，直到出现条件。

②条件变量类似 Event 对象的更高级版本。

③条件变量不能在进程之间共享。

④条件变量总是与某个锁关联。

⑤锁是条件变量的一部分，无须单独处理。

（5）事件：用 Event 类对象进行线程通信，其中一个线程发出事件信号，另一个线程等待信号。事件对象使用称为事件标志的内部标志。使用 set( )方法将其设置为 True，同样，使用 clear( )方法将其设置为 False。Event 类的 wait( )方法可以阻止线程，直到事件标志设置为 True。

（6）信号量：信号量是操作系统管理的抽象数据类型，为线程同步访问提供有限数量的资源。信号量是内部变量，反映对其关联资源的并发访问次数。信号量的值始终大于 0 且小于或等于可用资源总数。另外，信号量包含 acquire( )和 release( )方法。每次线程获取信号量同步资源时，信号量的值递减，同样，线程释放信号量同步资源时，信号量的值增加。

10. 用于等待直到其终止的线程方法是什么？

答：join( ) 方法。

11. 如何终止阻塞线程？

答：使用 thread.block( )或 thread.wait( )方法。

12. 信号量和有界信号量之间有什么区别？

答：有界信号量确保其当前值不超过它的初始值，对于信号量不存在这种情况。

13. RLock 是什么类型的锁？

答：重入锁。

14. 用于保护容量有限的资源（如数据库服务器）的同步方法是什么？

答：信号量。

15. 应使用哪种方法检测 Python 线程的状态？

答：isAlive( )方法。

16. 如何实现线程安全的全局锁保护？

答：借助全局解释器锁。

17. Python 中的哪个模块支持线程？

答：threading 模块。

18. 以下哪项陈述适用于 RLock？

（1）如果线程已拥有锁，则 acquire( )方法将递归级别递增 1，并立即返回。

（2）如果另一个线程拥有锁，则 acquire( )方法将阻塞，直到解锁。

答：这两种说法都正确。

19. 多线程对单处理器有何影响？

答：多线程会降低单处理器的性能。

20. Thread 类的哪两个方法在使用 threading 模块创建新线程时可对其进行更改？

答：__init__( )方法和 run( )方法。

21. 主线程始终是非后台线程。其余线程从其父线程继承后台线程性能。证明：

(1) 如果父线程是非后台线程，则子线程也是非后台线程。

(2) 如果父线程是后台线程，则子线程也是后台线程。

(3) 最后一个非后台线程终止时，自动终止所有后台线程。不需要显式终止后台线程。

答：

(1) 代码：

```
from threading import Thread, current_thread
def disp():
    print('Disp Function')

#证明主线程是非后台线程
mt = current_thread()
print(mt.getName())
print(mt.isDaemon())

#由于主线程正在执行 t1
#因此 t1 是主线程的子线程
t1 = Thread(target = disp)
print(t1.isDaemon())
t1.start()
```

输出：

```
MainThread
False
False
Disp Function
```

(2) 代码：

```
from threading import Thread, current_thread
def disp():
    print('Disp Function')
#如果 t1 是后台线程,那么 t2 也是后台线程
#如果 t1 是非后台线程,那么 t2 也是非后台线程
    t2 = Thread(target = show)
    print("Is t2 Daemon? ",t2.isDaemon())
    t2.start()
```

```
def show():
    print('show function')

#证明主线程是非后台线程
mt = current_thread()
print("Is main thread Daemon? ",mt.isDaemon())

#主线程正在执行 t1
#因此 t1 是主线程的子线程
t1 = Thread(target = disp)
print("Is t1 thread Daemon? ",t1.isDaemon())
t1.start()
```

输出:

```
Is main thread Daemon?    False
Is t1 thread Daemon?    False
Disp Function
>>>
Is t2 Daemon?    False
show function
```

(3) 代码:

```
import threading
import time

def work_for_t1():

    print('Starting of thread :', threading.current_thread().name)
    for i in range(0,11):
        print("Printing Number : ",i," \n")
        time.sleep(2)
    print('Finishing of thread :', threading.current_thread().
    name," \n")

t1 = threading.Thread(target =work_for_t1, name ='Thread1')
t1.setDaemon(True)
t1.start()
time.sleep(5)
print("Finishing thread: ", threading.current_thread().name," \n")
```

输出：

```
Starting of thread: Thread1
Printing Number : 0
Printing Number : 1
Printing Number : 2
Finishing thread: MainThread
```

第 6 章

错误、异常、测试与调试

> **引言**　Python 开发人员不仅要知道编程中可能遇到的各类错误和异常，还要知道它们的解决方法。本章介绍 Python 开发人员面临的各类错误和异常，并使用 Python 进行测试和调试。

知识结构

- 错误
  - 语法错误
  - 运行错误
  - 逻辑错误
- 异常
  - try 和 catch
  - 捕获通用异常
  - try…except…else 语句
  - try…except…finally 语句
  - try 和 finally
  - 引发异常
- 调试程序
- pdb 调试器
- 命令行调试器
- Python 的单元测试和测试驱动开发
- 测试级别
- pytest 概述
- unittest 模块
- 使用 unittest 和 pytest 定义多个测试用例
- unittest 模块中的主要 Assert 方法

完成本章的学习后,读者将掌握以下技能。
(1) 了解各类错误和异常;
(2) 捕获异常;
(3) 调试程序;
(4) 使用 unittest 模块创建单元测试。

6.1 错误

在开发软件时,经常需要反复运行代码以检查其是否正常工作。程序员有时会发现小问题并立即修复,有时会遇到不允许程序恢复并导致程序退出的问题,其无法捕捉或处理。代码中不让程序恢复并迫使其退出的问题称为错误。

错误分为 3 种,如图 6.1 所示。
(1) 语法错误;
(2) 运行错误;
(3) 逻辑错误。

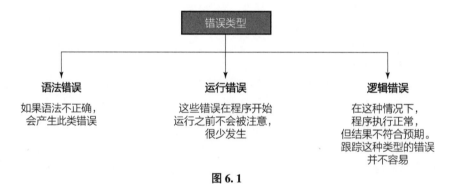

图 6.1

6.1.1 语法错误

编程语言由规则组成,这些规则通过组合命令、符号和语法来编写用于执行特定任务的可执行代码。当代码语法中出现错误时,开发人员就会遇到语法错误。
(1) 语法错误非常常见。
(2) 语法错误完全阻止整个程序运行。
(3) 语法错误可被轻易发现和修复。

示例 6.1
语法错误示例如下所示。

```
a = input("Please Enter a number: ")
b = input("Enter another number : ")
if (int(a)==int(b))
      print("both are equal")
else
      print("both are unequal")
```

如上所示，if 语句后缺少冒号（:）。

运行代码时，会出现错误消息框，如图 6.2 所示。

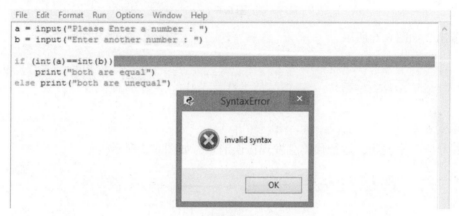

图 6.2

另外，还有一个语法错误为 else 语句后缺少冒号（:）。再次运行代码，得到图 6.3 所示结果。

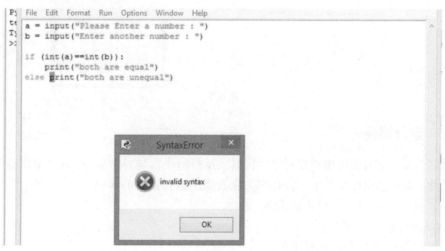

图 6.3

修复所有问题后，终于得到正确代码，并且运行正常，如图 6.4 所示。

```
File  Edit  Format  Run  Options  Window  Help
a = input("Please Enter a number : ")
b = input("Enter another number : ")
if (int(a)==int(b)):
    print("both are equal")
else:
    print("both are unequal")
```

图 6.4

输出 1 如图 6.5 所示。

```
Please Enter a number : 3
Enter another number : 3
both are equal
```

图 6.5

输出 2 如图 6.6 所示。

```
Please Enter a number : 6
Enter another number : 5
both are unequal
```

图 6.6

示例 6.2

括号未正常关闭，运行代码会发现错误消息，如图 6.7 所示。

```
import array as arr
a = arr.array('i', [9,3,2,90,65,23,45,77,76,57,0])
print("Array before Removing values"      ← 括号未关闭
for element in a:
    print(element, end = ' ')
print("\n removing 77")
a.remove(77)
print("removing 76")
a.remove(76)
print("removing 57")
a.remove(57)
print("removing 0")
a.remove(0)
print("Array after removing the elements")
for element in a:
    print(element, end = ' ')
```

SyntaxError: invalid syntax

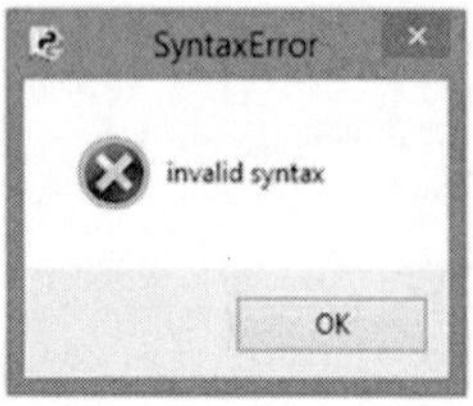

图 6.7

6.1.2 运行错误

顾名思义，运行错误出现在程序运行时。运行错误常与异常混淆，6.2 节将专门对此进行讨论。运行错误很少发生。将 0 作为除数就是运行错误的典例。

（1）运行错误在运行时才被检测。

（2）运行错误实例。

①无法找到数据或数据不存在；

②存在无效数据类型。

示例 6.3

观察下面的代码。

```
a = input("Please Enter father's age : ")
b = input("Enter Child's age : ")

print("age difference = ",(a-b))
```

代码似乎没问题，但运行时出现了图6.8所示的运行错误。

```
Please Enter father's age : 34
Enter Child's age : 4
Traceback (most recent call last):
  File "F:\2020 - BPB\Computer Science with Python - XI\Errors and Exceptions\code\errors.py", line 4, in <module>
    print("age difference = "+(a-b))
TypeError: unsupported operand type(s) for -: 'str' and 'str'
>>>
```

图6.8

此代码有何错误？

答：

输入为字符串类型，相减之前，需要将其转换为整型。

修改后代码如下所示。

```
a = input("Please Enter father's age : ")
b = input("Enter Child's age : ")

print("age difference = ",(int(a)-int(b)))
```

输出结果如图6.9所示。

```
Please Enter father's age : 34
Enter Child's age : 4
age difference =  30
```

图6.9

示例6.4

下面的代码提示用户输入两个数字，然后将两数相除并显示输出结果。

```
x = int(input("x = "))
y = int(input("y = "))
print(x/y)
```

输出1：

```
x = 8
y = 4
2.0
```

如果用户输入 x=8 和 y=0，则输出结果是什么？

输出2：

```
x = 8
y = 0
Traceback (most recent call last):
  , line 3, in <module>
    print(x/y)
ZeroDivisionError: division by zero
```

6.1.3 逻辑错误

逻辑错误不会在代码运行时引发问题。只有运行代码后，才能发现结果与预期存在出入。逻辑错误通常在测试代码逻辑功能时出现。

查看下面的代码。

```
x = int(input("x = "))
y = int(input("y = "))
print(" The sum of x and y = ",x*y)
```

输出 1:

令 x = 2，y = 2，首次运行该代码。

```
x = 2
y = 2
The sum of x and y =   4
```

似乎并未出错。令 x = 15，y = 2，再次运行该代码。

输出 2:

```
x = 15
y = 2
The sum of x and y =   30
```

预期输出为 17，但结果为 30。逻辑关系存在问题。文本显示为 x 和 y 的和，而输出为 x 和 y 的乘积。因此，要么文本错误，要么输出错误。

> （1）语义错误。
>
> 语义错误是指有意义数据提供不足的错误。它究竟是逻辑错误还是语法错误，学界众说纷纭。逻辑错误由错误的逻辑引起，而语义错误则产生无意义的输出。例如，计算课堂缺勤人数百分比，结果误将（缺勤人数/总人数）乘以 10 而非 100。编译期间不会检测到语义错误。因此，许多程序员认为语义错误和逻辑错误相同。
>
> （2）设计错误。
>
> 设计错误的产生归因于缺乏对客户需求的理解。假设代码正确且没有漏洞和缺陷（bug），完全符合自身期望，然而软件并不完全符合客户期望。这有时是因为沟通问题，即甲方未能详细表达需求，或者乙方没有正确记录需求细节。这就是所谓设计错误。

6.2 异常

异常是程序执行时中断程序运行的事件。程序运行时也会发生错误，即运行错误。因此，两者常被混淆，有时人们也认为异常与运行错误相同。

异常与运行错误的不同在于异常可被处理，并非无条件中断程序。事实上，错误被认为是未经检查的异常。

下面介绍 Python 中的异常。用 Python 编码时，常会遇到语法错误和异常。

语法错误在编译时识别，如果未正确遵循编码语法规则，就会发生语法错误。可以对运行时发生的异常进行处理，以便程序继续运行。

6.2.1 try 和 catch

再次讨论示例 6.4 中的逻辑。

```
x = int(input("x = "))
y = int(input("y = "))
print(x/y)
```

用户将 y 赋值为 0 之前，代码运行正常。赋值后产生错误，如下所示。

```
ZeroDivisionError: division by zero
```

虽然 ZeroDivisionError 是非致命错误，但会影响程序运行。不过可以对其进行处理以防程序执行中断。使用 try - except 块处理。当 y = 0 时，语句 divide = x/y 引发异常。使用 try - except 块的方法如下。

步骤 1：将产生错误的代码放在 try 块中。代码如下所示。

```
x = int(input("Enter first number : "))
y = int(input("enter second number : "))
try:
    divide = x/y
    print("try block over")
    print("divide = ",divide)
```

步骤 2：定义 except 块。

在 except 块中，定义发生错误时需执行的语句。代码如下所示。

```
except ZeroDivisionError:
print("OOPS exception occured due to Zero division error")
print(" * * * * THE END * * * * ")
```

代码：

```
x = int(input("Enter first number : "))
y = int(input("enter second number : "))
try:
    divide =  x/y
    print("try block over")
    print("divide = ",divide)
except ZeroDivisionError:
    print("OOPS exception occured due to Zero division error")
    print("****THE END****")
```

（1）执行 try 块中的语句。

（2）如果未发生错误，则跳过异常块；如果发生错误，并且其类型与以 except 关键字命名的异常名称（本示例中为 ZeroDivisionError）匹配，则执行 except 块。

输出：

实例 1：一切正常；执行 try 块，跳过 except 块。

```
Enter first number : 1
enter second number : 2
try blockover
divide = 0.5
****THE END****
```

实例 2：发生异常；跳过 try 块中的指令。该错误与 except 块中的错误匹配并执行。

```
Enter first number : 1
enter second number : 0
OOPS exception occured due to Zero division error
****THE END****
```

示例 6.5

如果错误与 except 块中的异常名称不匹配会怎样？

如果发生的异常与 except 子句中的异常名称不匹配，则将其作为未处理异常，执行中断并给出提示。

代码：

```
a = int(input("Enter first number : "))
b = input("enter second number : ")
try:
        divide =  a/b
        print("try block over")
        print("divide = ",divide)
```

```
except  ZeroDivisionError:
        print("OOPS exception occurred due to Zero division error")
        print(" **** THE END **** ")
```

输出:

```
Enter first number : 1
enter second number : 2
Traceback (most recent call last):
  File "F://Errors and Exceptions/code/trycatch.py", line 4, in <module>
    divide = a/b
TypeError: unsupported operand type(s) for /: 'int' and 'str'
```

示例 6.5 中出现了 TypeError，而 except 块只定义处理 ZeroDivisionError。因为未找到 TypeError 的处理程序，所以该错误未处理，代码也随之停止执行。

下面代码中的 TypeError 也作为异常名添至 except 关键字前面。

代码:

```
a = int(input("Enter first number : "))
b = input("enter second number : ")
try:
        divide = a/b
        print("try block over")
        print("divide = ",divide)
except  (ZeroDivisionError,TypeError)  :
        print("OOPS exception occurred due to Exception")
        print(" **** THE END **** ")
```

输出:

```
Enter first number : 1
enter second number : 2
OOPS exception occurred due to Exception
**** THE END ****
```

由此可见，except 块处理 TypeError 的优先级与 ZeroDivisionError 相同。

try 块可有多个 except 为不同异常指定处理程序，但最多只执行一个处理程序。处理程序只处理与之关联的 try 块中发生的异常，不处理同一 try 块其他处理程序中存在的异常。

6.2.2 捕获通用异常

因为有时无法确定出现何种异常，所以希望捕获通用异常并获取更多相关细节。下面为获取信息的简单方式。

代码:

```
a = int(input("Enter first number : "))
b = input("enter second number : ")
```

```
try:
    divide = a/b
    print("try block over")
    print("divide = ",divide)
except TypeError as e:
    print(str(e))
```

输出:

```
Enter first number : 1
enter second number : 2
unsupported operand type(s) for /: 'int' and 'str'
```

这是对错误的简单解释,但处理数千行代码时效果不佳。有时希望获得更多错误信息,以便更易查明错误。为此,首先导入 sys 模块。其为 Python 的重要模块,提供对解释器用到的重要变量和函数的访问。代码如下所示。

```
import sys
```

然后使用 sys.exc_info() 函数获取异常的详细信息。

sys.exc_info() 函数返回由 3 个值组成的元组。

这 3 个值提供了正在处理的异常信息。

(1) type:处理的异常类型。

(2) value:获取异常参数。

(3) traceback:获取 traceback 对象,该对象在最初发生异常时封装调用堆栈。

示例 **6.6**

编程获取发生的异常类型。

代码:

```
import sys
a = int(input("Enter first number : "))
b = input("enter second number : ")
try:
    divide = a/b
    print("try block over")
    print("divide = ",divide)
#使用 sys.exc_info()函数获取更多信息
except:
    print("OOPS exception occurred due to ",sys.exc_info()[0])
    print("****THE END****")
```

输出:

```
Enter first number : 1
enter second number : 2
OOPS exception occured due to   <class 'TypeError'>
****THE END****
```

示例 6.7

编写程序获取发生异常的 value。

代码:

```
import sys
a = int(input("Enter first number : "))
b = input("enter second number : ")
try:
    divide = a/b
    print("try block over")
    print("divide = ",divide)
#使用sys.exc_info()函数获取更多信息
except:
    print("OOPS exception occurred due to ",sys.exc_info()[1])
    print(" **** THE END **** ")
```

输出:

```
Enter first number : 1
enter second number : 6
OOPS exception occurred due to unsupported operand type(s) for /: 'int' and 'str'
**** THE END ****
```

示例 6.8

编程获取发生异常的 traceback。

代码:

```
import sys
a = int(input("Enter first number : "))
b = input("enter second number : ")
try:
    divide = a/b
    print("try block over")
    print("divide = ",divide)
#使用sys.exc_info()函数获取更多信息
except:
    print("OOPS exception occurred due to ",sys.exc_info()[2])
    print(" **** THE END **** ")
```

输出:

```
Enter first number : 1
enter second number : 2
OOPS exception occurred due to  <traceback object at 0x010BF288 >
**** THE END ****
```

示例 6.9
编程获取发生异常的线路信息表。
代码：

```
import sys
a = int(input("Enter first number : "))
b = input("enter second number : ")
try:
    divide = a/b
    print("try block over")
    print("divide = ",divide)
#使用 sys.exc_info()函数获取更多信息
except:
    print("OOPS exception occurred due to line no.",sys.exc_info()[2].
    tb_lineno)
    print("****THE END****")
```

输出：

```
Enter first number : 2
enter second number : 2
OOPS exception occurred due to line no.5
****THE END****
```

6.2.3　try…except…else 语句

try…except 语句附带一个可选 else 子句。该子句需要放在所有 except 子句后。else 子句用于即使 try 子句未引发异常，也必须执行某些代码的情况。

实例 1：发生异常，else 子句未执行。
代码：

```
import sys
a = int(input("Enter first number : "))
b = input("enter second number : ")
try:
    divide = a/b
    print("divide = ",divide)
except:
    print("OOPS exception occurred due to line no.",sys.exc_info()[2].
    tb_lineno)
else:
    print("try over successfully")
    print("****THE END****")
```

输出：

```
Enter first number : 1
enter second number : 3
OOPS exception occurred due to line no. 5
**** THE END ****
```

实例2：未发生异常，else 子句执行。

代码：

```
a = int(input("Enter first number : "))
b = int(input("enter second number : "))
try:
    divide = a/b
    print("divide = ",divide)
except:
    print("OOPS exception occurred due to line no.",sys.exc_info()[2].tb_lineno)
else:
    print("try over successfully")
    print(" **** THE END **** ")
```

输出：

```
Enter first number : 2
enter second number : 6
divide =  0.3333333333333333
try over successfully
**** THE END ****
```

在 else 子句而不是 try 子句中增加代码，可避免意外捕获由 try…except 语句保护的代码引发的异常。

6.2.4　try…except…finally 语句

try…except…finally 语句的使用方法如下。

代码：

```
def just_having_fun():
    try:
        print(4)
    except TypeError as e:
        print(str(e))
    finally:
        print("Ok Bye")

just_having_fun()
```

输出：

```
4
Ok Bye
>>>
```

6.2.5　try 和 finally

将 try 块与 finally 块配合使用。在 finally 块中放置必须执行的代码，不必考虑 try 块是否抛出异常。

语法格式如下所示。

```
try:
# 执行语句
    ----------------------
    ----------------------
finally:
# 经常执行的语句
    ----------------------
    ----------------------
```

6.2.6　引发异常

使用 raise 子句引发异常。语法格式如下所示。

```
raise Exception_class
```

（1）raise 语句引发异常。
（2）raise 语句后跟异常类名。

代码：

```
try:
    voter_age = int(input("Enter the age of the voter : "))
    if voter_age < 18:
        raise ValueError;
    else:
        print("You can vote")
except ValueError:
    print("You are not allowed to vote")
```

输出 1：

```
Enter the age of the voter : 6
You are not allowed to vote
```

输出 2：

```
Enter the age of the voter : 19
You can vote
```

图 6.10 所示为编程时常见的标准异常。

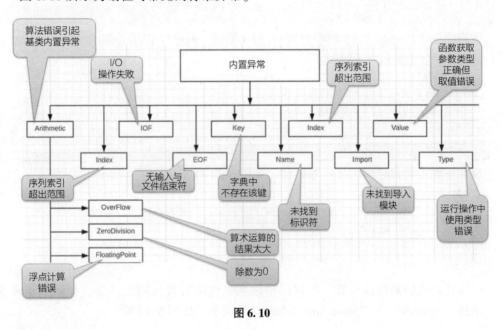

图 6.10

6.3 调试程序

学习调试程序对成为一名优秀的程序开发人员非常重要。注意生成错误、其生成行以及错误类型，就能轻易找出错误。手动调试时，程序开发人员解决问题的能力完全取决于其对调试程序的理解。至此读者已经学习了如何使用异常处理可疑代码，以及获取出错原因的更多信息。本节介绍找出代码的错误位置的基本技术。

小错误可以被轻易发现。例如，首次阅读下面的代码时，似乎一切正常。

```
# 如何调试程序
class Student:
    def __init__(self, name, email, department, age):
        self.name = name
        self.email = email
        self.department = department
        self.age = age

    def print_student(self):
        print("Name : ",self.name)
```

```
        print("Email : ",self.email)
        print("Department : ",self.department)
        print("Age : ",self.age)

s1 = Student("Alex","alex@company_name.com","Electronics",18)
s1.print_stdent()
```

该程序创建 Student 类的实例并打印各种变量值。创建对象 s1 并调用 print_student()。运行该程序时，会显示以下错误。

```
Traceback (most recent call last):
  File " identifyerrors.py", line 16, in <module>
    s1.print_stdent()
AttributeError: 'Student' object has no attribute 'print_stdent'
```

现在仔细查看该错误，显示 line 16，如图 6.11 所示。

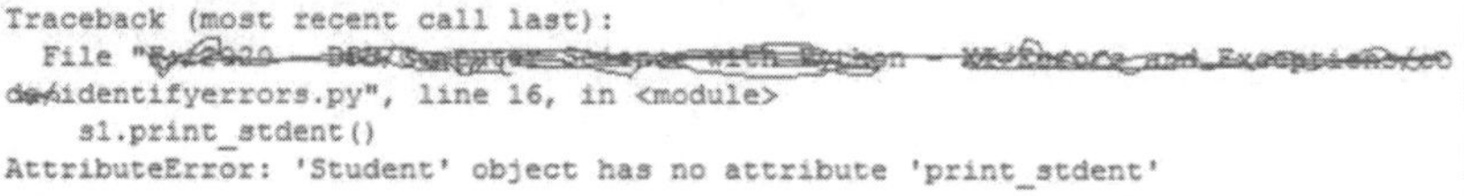

图 6.11

小段代码能轻易数清行数，但代码量巨大，代码行数也随之增加。打开 Python 文件，选择 "Options" → "Show Line Numbers" 选项，如图 6.12 所示。

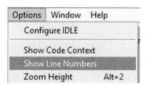

图 6.12

在代码左侧显示行号，如图 6.13 所示。

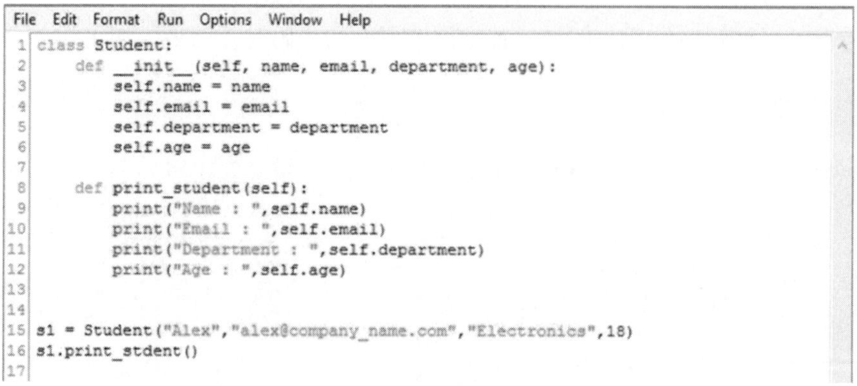

图 6.13

由此得到实际引发错误的代码行（图 6.11）。

错误信息进一步解释为 AttributeError：'Student' object has no attribute'print_stdent'，其指向单词 print_stdent。

仔细查看，发现存在拼写错误。函数名为 print_student()，但作为对象调用函数 print_stdent() 不存在。因此，拼写上的小错误就能终止整个程序运行。

更改之后，代码如下所示。

```
class Student:
    def __init__(self, name, email, department, age):
        self.name = name
        self.email = email
        self.department = department
        self.age = age

    def print_student(self):
        print("Name : ",self.name)
        print("Email : ",self.email)
        print("Department : ",self.department)
        print("Age : ",self.age)

s1 = Student("Alex","alex@company_name.com","Electronics",18)
s1.print_student()
```

输出结果如下所示。

```
Name : Alex
Email : alex@company_name.com
Department : Electronics
Age : 18
>>>
```

一种调试方法为输出变量值。查看如下示例。其中不存在语法错误，因此编译时未检测到错误，但输出结果与实际计算结果不匹配。代码从用户输入取两个数，第 1 个数与 2 相乘，第 2 个数与 3 相乘，然后将两个结果相加。因此，如果取两个数 2 和 4，最后结果为 4 + 12 = 16。然而，事实并非如此。

```
a = input("enter a number : ")
b = input("enter one more number : ")
c = a*2
d = b*3
print(c+d)
```

上面代码的输出结果如下所示。

```
enter a number : 2
enter one more number : 4
22444
```

期望结果为 16，但实际结果为 22444，表明存在逻辑错误。这时通过输出变量值寻找问题原因，如下所示。

```
a = input("enter a number : ")
print("value of first number a is ",a)
b = input("enter one more number : ")
print("value of second number b is ",b)
c = a*2
print("c is a multiplied by 2 i.e. ",c)
d = b*3
print("d is b multiplied by 3 i.e. ",d)
print("c + d = ",c+d)
```

观察输出结果并分析错误原因，如下所示。

```
enter a number : 1
value of first number a is  1
enter one more number : 2
value of second number b is  2
c is a multiplied by 2 i.e.  11
d is b multiplied by 3 i.e.  222
c + d =  11222
>>>
```

$a=1$，所以 $c=a*2=2$。

$b=2$，所以 $d=b*3=6$。

分析输出结果，可知 Python 将这些数视为字符串，然后执行了字符串连接操作。仔细观察，发现输入值未先转换成整数。

更正后代码如下所示。

```
a = int(input("enter a number : "))
print("value of first number a is ",a)
b = int(input("enter one more number : "))
print("value of second number b is ",b)
c = a*2
print("c is a multiplied by 2 i.e. ",c)
d = b*3
print("d is b multiplied by 3 i.e. ",d)
print("c + d = ",c+d)
```

输出 1：

```
enter a number : 1
value of first number a is  1
```

```
enter one more number : 2
value of second number b is  2
c is a multiplied by 2 i.e.  2
d is b multiplied by 3 i.e.  6
c + d =  8
```

输出 2：

```
enter a number : 2
value of first number a is  2
enter one more number : 4
value of second number b is  4
c is a multiplied by 2 i.e.  4
d is b multiplied by 3 i.e.  12
c + d =  16
```

另一种调试技术为代码跟踪，即每执行一行，便查看代码实际执行各种函数的情况。常使用调试器实现代码跟踪。

6.4 Python 调试器

pdb（Python 调试器）为命令行工具。

（1）安装 Python 时安装在操作系统中的 Python 模块。

（2）调试 Python 程序。

（3）学习 pdb 很重要，在实时场景中，需要在服务器上运行脚本，而在服务器上无法运行 Python IDE。可以通过 pdb 协助调试 Python 代码。

（4）允许设置断点，以便：

①检查变量；

②逐行运行；

③执行时显示代码。

（5）pdb 中有两个最重要语句。

①导入语句：import pdb。

②pdb.set_trace()：在检查代码处插入断点。

代码如下所示。

```
import pdb
a = int(input("enter a number : "))
b = int(input("enter one more number : "))
pdb.set_trace()
c = a*2
d = b*3
pdb.set_trace()
print("c + d = ",c+d)
```

6.5 命令行调试器

表 6.1 所示为调试时可使用的命令。图 6.14 演示了各命令的用法。

表 6.1

命令	说明
'n'	执行下一行代码
'c'	继续执行代码/完成执行
'l'	列出当前执行语句的前后 3 行代码
's'	插入函数并逐行执行
'b'	列出所有断点
'b[int]'	设置行号不同的断点
'b[func]'	在函数名处断开
'cl'	清除所有断点
'cl[int]'	清除指定行号断点
'p'	输出

```
enter a number : 4
enter one more number : 2
> f:\2020 - bpb\computer science with python - xi\errors and exceptions\code\identifyerrors.py(5)<module>()
-> c = a*2
(Pdb)
(Pdb) l
  1     import pdb
  2     a = int(input("enter a number : "))
  3     b = int(input("enter one more number : "))
  4     pdb.set_trace()
  5  -> c = a*2
  6     d = b*3
  7     pdb.set_trace()
  8     print("c + d = ",c+d)
[EOF]
(Pdb) p(a)
4
(Pdb) p(b)
2
(Pdb) n
> f:\2020 - bpb\computer science with python - xi\errors and exceptions\code\identifyerrors.py(6)<module>()
-> d = b*3
(Pdb) p(d)
*** NameError: name 'd' is not defined
(Pdb) n
> f:\2020 - bpb\computer science with python - xi\errors and exceptions\code\identifyerrors.py(7)<module>()
-> pdb.set_trace()
(Pdb) p(d)
6
(Pdb) cl
Clear all breaks? y
```

图 6.14

6.6 Python 的单元测试和测试驱动开发

本节介绍单元测试的必要性和测试驱动开发。

1. 单元测试

单元测试是程序软件测试级的第一个,应尽早捕获错误以防错误出现在最终产品中。单元测试的作用如下。

(1) 在很大程度上减少现有代码的漏洞和缺陷,并简化添加新特性的过程。

(2) 提供详细报告,说明测试运行过程及其是否符合预期。

(3) 显著降低产品的变更成本,能在软件开发生命周期早期捕获并修复漏洞和缺陷,从而节省程序开发人员大量的时间精力。在软件开发后期阶段修复错误较为困难。软件错误会影响业务,导致公司声誉受损,顾客将选择其他产品。

(4) 支持更快调试。

(5) 有助于开发设计更好的程序。

2. 单元测试的目的

单元测试的目的如下。

(1) 测试代码的单个功能。

(2) 通过识别错误以确保软件基本单元无误。

(3) 为每个单元的所有功能编写所有类型的阳性和阴性测试。

(4) 所有测试均需要在开发环境而非生产环境中执行,以便不限时间和次数地运行。

6.7 测试级别

本节介绍不同的测试级别——单元测试、集成测试、系统测试、性能测试,以及 Python 的单元测试和 Python 编程中的常见错误。

1)单元测试

软件测试在不同级别进行,这些级别能确保及时发现错误,并阻止错误进入最终产品。单元测试为软件测试第 1 级,在源代码基本单元中进行。单元测试检验其是否可以进行相互集成。因此,单元测试是最低级别的测试(对函数、子例程和类进行测试,以检查其能否达到预期结果),也是最全面的测试,可测试功能中所有阳性和阴性测试用例。

2)集成测试

将单元集成在一起,并针对集成组件进行集成测试。

3)系统测试

系统测试在子系统集合的系统外部接口上进行。

4）性能测试

完成单元、集成和系统级别的测试后，对软件进行子系统级别和系统级别的性能测试。性能测试期间，测试软件以保证时间和资源使用（如内存、CPU、磁盘使用）在接受范围内。

5）Python 的单元测试

软件开发的理想方法是先理解问题，创建测试用例，然后编写代码。虽然未要求必须遵循该方法，但毫无疑问，这是软件开发的最佳方式。

6）Python 编程中的常见错误

Python 编程中的常见错误如下。

（1）变量名拼写错误。

（2）变量初始化或重新初始化失败。

（3）程序员混淆值相等和对象相等。例如，a == b 和 id(a) == id(b) 并不相同。

（4）逻辑错误。

下面将介绍 Python 中最流行的两个测试框架：pytest 和 unittest。

> **Assertions（断言）**
>
> 使用 pytest 和 unittest 框架进行单元测试时会用到 Assertions。
>
> Assertions 是带有表达式的测试。如果表达式的计算结果为 True，则表示测试通过；如果表达式的计算结果为 False，则表示测试失败。
>
> Assertions 由 assert 语句协助执行。Python 计算与 assert 语句相关的表达式结果。如果表达式结果为 False，则 Python 生成 AssertionError。
>
> assert 语句的语法格式如下所示。
>
> ```
> assert Expression(Arguments)
> ```

6.8 pytest 概述

pytest 实际上是用于单元测试的 Python 框架，允许创建测试、模块和固定程序。相比其他单元测试框架，pytest 更简单、方便。本节介绍以下内容。

（1）安装 pytest。

（2）Python 内置 assert 语句的简化单元测试实现。

（3）使用命令行参数过滤将执行哪些测试顺序。

使用 pip 安装 pytest 的操作如下。

在命令行中执行以下指令：

```
pip install pytest
```

检查其是否安装，转到 Python Shell 并写入：

```
import pytest
```

检查 pytest 是否已被安装,如图 6.15 所示。

```
>>> import pytest
>>>
```

图 6.15

未生成错误,表示 pytest 安装成功。

下面开始第一个程序。

步骤1:编写代码。

创建文件来定义用于测试的函数。本示例编写定义用于加、减、乘和除的函数。代码如下所示。

```
def add(x,y):
    return x + y

def subtract(x,y):
    return x - y

def multiply(x,y):
    return x * y

def divide(x,y):
    if(y == 0):
        str1 = "invalid denominator"
        return str1
    else:
        return x/y
```

保存文件。本示例文件以"arith_funcs.py"命名并保存至目录"F:\testing"下。

步骤2:编写文件用于测试代码。

创建另一个文件用于测试上述函数。

文件名称应以 test_ 开头。

在本示例中,文件名为"test_arith_funcs.py"。

(1) 导入测试文件。代码如下所示。

```
import arith_funcs
```

(2) 定义测试。

为测试编写的 Python 函数需要在函数名增加前缀 test。每个测试函数都调用其待测函数并验证操作结果,为此使用 assert 关键字。例如,"arith_funcs.py"文件中的 add() 函数必须提供两个数的加法运算。因此,编写 assert arith_funcs.add(10,20) = 30,即定义调用"arith_funcs.py"文件中的add()函数并传递参数10与20，其输出应为30，否则生成错误。其他函数的assert语句也以相同方式定义。代码如下所示。

```
import arith_funcs

def test_add():
    assert arith_funcs.add(10,20) == 30

def test_subtract():
    assert arith_funcs.subtract(10,20) == -10

def test_multiply():
    assert arith_funcs.multiply(10,20) == 200

def test_divide():
    assert arith_funcs.divide(10,0) == "invalid denominator"
    assert arith_funcs.divide(10,5) == 2
```

（3）保存文件。

文件命名为"test_arith_funcs.py"，保存至目录"F:\testing"下。

（4）执行测试文件。

①打开命令提示符窗口。

②输入pytest指令。

执行该文件的语法格式如下所示。

```
pytest test_pythonfile.py
```

本示例为test_pythonfile.py = F:\testing\test_arith_funcs.py。

在命令提示符窗口中输入以下命令：

```
pytest F:\testing\test_arith_funcs.py
```

所有测试均正常执行，如图6.16所示。

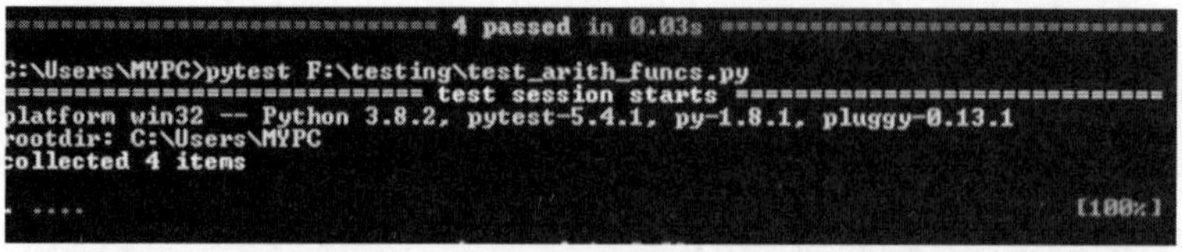

图6.16

现在看看如果代码存在错误，会出现何种情况。为此将代码稍微修改。源文件中已定义，在divide()函数中，如果分母为0，则应返回字符串"invalid denominator"。假设在测试文件中放置Assertions，如果分母为0，则返回的字符串应为"zero not allowed as denominator"。因为两个字符串不匹配，所以测试失败。下面进行测试。代码如下所示。

"F:\testing\test_arith_funcs.py"文件的代码如下所示。

```
import arith_funcs

def test_add():
    assert arith_funcs.add(10,20) == 30

def test_subtract():
    assert arith_funcs.subtract(10,20) == -10

def test_multiply():
    assert arith_funcs.multiply(10,20) == 200

def test_divide():
    assert arith_funcs.divide(10,0) == "zero not allowed as denominator"
    assert arith_funcs.divide(10,5) == 2
```

输出结果如图 6.17 所示。

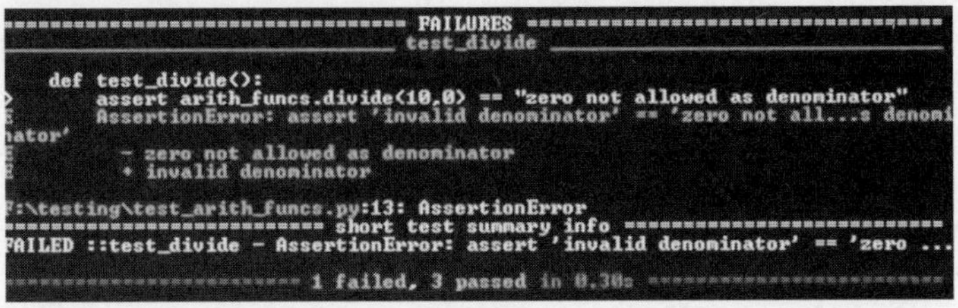

图 6.17

6.9 unittest 模块

6.8 节介绍了如何使用 pytest 进行单元测试。现在介绍如何使用 Python 库的 unittest 模块，其组成如下。

（1）构成测试用例基础的核心框架类。
（2）用于运行测试和报告结果的实用程序类。
（3）用于执行单元测试的 Testcase 类。

1. 测试方法的编写规则

（1）如果测试方法名称没有以 test 开头，其将被忽略。因此，第 1 条规则是以 test 一词开始测试名称。
（2）所有测试方法将只有一个参数——self。
（3）self 用于调用所有内置 Assertions 方法。

现在通过相同的"arith_funcs.py"文件示例学习使用 unittest 模块。代码如下所示。

```
def add(x,y):
    return x + y

def subtract(x,y):
    return x - y
```

```python
def multiply(x,y):
    return x * y

def divide(x,y):
    if(y ==0):
        str1 = "invalid denominator"
        return str1
    else:
        return x/y
```

2. 使用 unittest 模块需要遵循的步骤

步骤 1：导入 unittest 模块。

在"test_arith_funcs. py"文件中导入 unittest 模块。测试代码如下所示。

```python
import arith_funcs
import unittest

class TestArithFuncs(unittest.TestCase):
#定义测试方法

    def test_add(self):
        self.assertEqual(30,arith_funcs.add(10,20))
    def test_subtract(self):
        self.assertEqual(-10,arith_funcs.subtract(10,20))
    def test_multiply(self):
        self.assertEqual(200,arith_funcs.multiply(10,20))
    def test_divide(self):
        self.assertEqual("zero not allowed as denominator",arith_funcs.divide(10,0))
        self.assertEqual(2,arith_funcs.divide(10,5))

unittest.main()
```

输出结果如图 6.18 所示。

```
= RESTART: F:\2020 - BPB\Python for Undergraduates\Testing and Debugging\Code\te
st_arith_funcs.py
.F..
======================================================================
FAIL: test_divide (__main__.TestArithFuncs)
----------------------------------------------------------------------
Traceback (most recent call last):
  File "F:\2020 - BPB\Python for Undergraduates\Testing and Debugging\Code\test_
arith_funcs.py", line 17, in test_divide
    self.assertEqual("zero not allowed as denominator",arith_funcs.divide(10,0))
AssertionError: 'zero not allowed as denominator' != 'invalid denominator'
- zero not allowed as denominator
+ invalid denominator


----------------------------------------------------------------------
Ran 4 tests in 0.010s

FAILED (failures=1)
>>>
```

图 6.18

由此可见，执行了 4 个测试，其中有 1 个失败。在上述代码中，如果将一个数除以 0，应返回字符串"invalid denominator"，而测试用例 Assertions 返回的字符串的值应为"zero not allowed as denominator"。因为两个值不同，所以该测试用例失败。

步骤 2：修改"arith_funcs.py"文件，并再次执行测试。代码如下所示。

```
def add(x,y):
    return x+y

def subtract(x,y):
    return x-y

def multiply(x,y):
    return x*y

def divide(x,y):
    if(y==0):
        //变更代码
        str1 = "zero not allowed as denominator"
        return str1
    else:
        return x/y
```

输出结果如图 6.19 所示。

```
Ran 4 tests in 0.074s

OK
```

图 6.19

6.10 使用 unittest 和 pytest 定义多个测试用例

使用 unittest 和 pytest 定义多个测试用例并执行。

（1）使用 unittest 定义多个测试用例。代码如下所示。

```
import arith_funcs
import unittest

class TestArithFuncs(unittest.TestCase):
    #定义测试方法

    def test_add(self):
        vals = [(10,20),(20,78),(60,54),(13,25)]
        ans = [30,98,114,38]
        for i in range(len(vals)):
            self.assertEqual(ans[i],arith_funcs.add(vals[i][0],vals[i][1]))

    def test_subtract(self):
        self.assertEqual(-10,arith_funcs.subtract(10,20))
```

```
    def test_multiply(self):
        self.assertEqual(200,arith_funcs.multiply(10,20))

    def test_divide(self):
        self.assertEqual("zero not allowed as denominator",arith_
        funcs.divide(10,0))
        self.assertEqual(2,arith_funcs.divide(10,5))

unittest.main()
```

因此，可在 test_add() 函数中观察到已使用列表创建多个用例。输出结果如图 6.20 所示。

```
....
-------------------------------------------------------------
Ran 4 tests in 0.023s

OK
```

图 6.20

（2）使用 pytest 定义多个测试用例。

"arith_func.py" 文件的代码如下所示。

```
def add(x,y):
    return x + y

def subtract(x,y):
    return x - y

def multiply(x,y):
    return x * y

def divide(x,y):
    if(y == 0):
        str1 = "invalid denominator"
        return str1
    else:
        return x/y
```

"test_arith_func.py" 文件的代码如下所示。

```
import arith_funcs

def test_add():
    vals = [(10,20),(20,78),(60,54),(13,25)]
    ans = [30,98,114,38]
    for i in range(len(vals)):
        assert arith_funcs.add(vals[i][0],vals[i][1]) == ans[i]
```

```
def test_subtract():
    assert arith_funcs.subtract(10,20) == -10

def test_multiply():
    assert arith_funcs.multiply(10,20) == 200

def test_divide():
    assert arith_funcs.divide(10,0) == "invalid denominator"
    assert arith_funcs.divide(10,5) == 2
```

输出结果如图 6.21 所示。

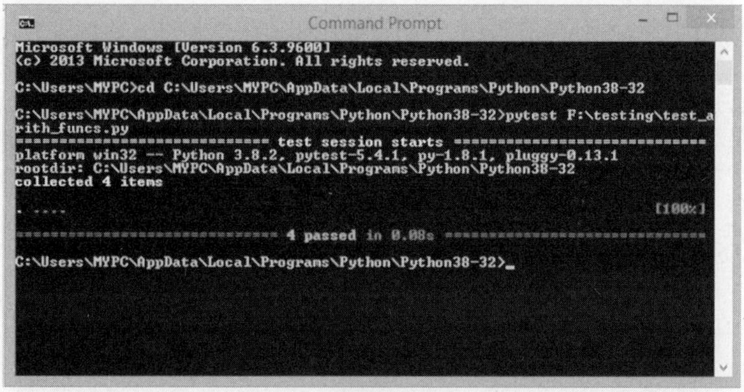

图 6.21

6.11 unittest 模块中的主要 Assert 方法

表 6.2 所示为各种 Assertion 方法及其说明。

表 6.2

Assertion 方法	说明
self. assertEqual (expected _ Result，actual _ Result，[Message])	预期结果与实际结果相同时，测试通过
self. assertNotEqual(expected_Result，actual_Result，[Message])	预期结果与实际结果不同时，测试通过
self. assertAlmostEqual (expected _ Result，actual _ Result，[Message])	检查是否超过预期，实际结果在四舍五入到小数点后几乎相等
self. assertTrue(Condition，[Message])	如果条件为 True，测试通过
self. assertFalse(Condition，[Message])	如果条件为 False，测试通过
self. assertRaises (exception，functionName，parameter，…，parameter)	测试给定参数调用函数 functionName ( ) 是否引发给定异常

小结

测试是软件开发的重要部分。程序开发人员必须在程序开发过程中不断测试代码,以便开发出无漏洞和缺陷的应用程序。学习错误和异常的类型非常重要,以便确定程序存在的问题类型。一旦发现问题,便可轻松解决。

简答题

1. 解释语法错误、运行错误和逻辑错误的区别。

答:表6.3对语法错误、运行错误和逻辑错误的区别做了总结。

表6.3

总结	语法错误	运行错误	逻辑错误
诱因	语法格式、字符或标记的序列错误	语句错误	执行逻辑错误
结果	无法编译	无法执行	输出错误结果
错误消息	显示	显示	不显示
检测	解析代码期间	运行期间	测试逻辑错误期间

2. 借助图表,解释何时使用 try 块。

答:如果怀疑代码中某些部分会抛出异常,则将其放入 try 块。try 块后必须紧跟包含代码的 except 块,如果 try 块中的代码抛出异常,则执行 except 块中的代码。

简单的 try – except 块如图6.22所示。

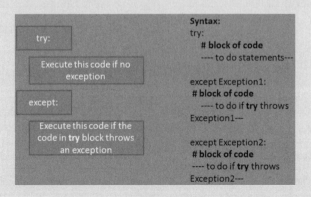

图6.22

try…except…else 语句如图 6.23 所示。

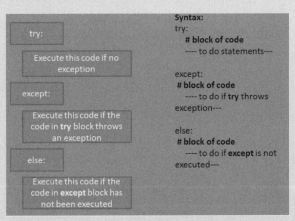

图 6.23

（1）不提及异常名称的 except 块。可以选择在 except 块中不指定异常名称。
（2）通过 try – except 块指定 else 块，如果 try 块未引发异常，则执行该语句。
（3）不抛出异常的语句必须放在 else 块中。

3. 如何声明多个异常？

答：

语法格式如下所示。

```
try:
    #代码块
    except( <Exception1 >, <Exception2 >, <Exception3 >,…, <Exceptionn >)
    #代码块
else:
    #代码块

The finally block:
#可以使用 try 块和 finally 块
```

4. 如何引发异常？

答： 使用 raise 子句引发异常。

语法格式如下所示。

```
raise Exception_class < value >
```

（1）raise 语句用于引发异常。
（2）raise 语句后跟 Exception 类名称。

5. 判断错误类型。

（1）缺少冒号；

(2) 除数为 0；
(3) 加法代替减法；
(4) 选择语句和循环语句格式错误；
(5) 试图打开不存在文件；
(6) 缺少圆括号、方括号和花括号；
(7) 显示错误消息；
(8) 从错误文件中获取数据。

答：
(1) 语法错误；
(2) 运行错误；
(3) 逻辑错误；
(4) 语法错误；
(5) 运行错误；
(6) 语法错误；
(7) 逻辑错误；
(8) 逻辑错误。

6. 区分语法错误和运行错误。

答：
语法错误和运行错误的区别如表 6.4 所示。

表 6.4

语法错误	运行错误
由语法错误陈述引起	语法陈述正确
解析代码时发生错误	代码被解析为语法正确后，执行代码时发生错误
示例： • 缺少操作符 • 一行中的两个运算符间没有分号 • 括号匹配错误	示例： • 变量类型或大小错误 • 变量不存在 • 超出范围等

7. 以下横线处应填入什么内容？

(1) _____ 错误发生时，语法正确但代码执行时发生错误。

(2) _____ 暂停解释器运行，报告错误，终止执行程序。

(3) 对于 _____ 错误，解释器执行程序，但遇见错误时停止，并将其报告为异常。

(4) 对于＿＿＿＿错误，解释器继续运行程序且不报告错误。

答：

(1) 运行；

(2) 语法错误；

(3) 运行；

(4) 逻辑/语法。

8. 以下代码将产生什么错误？

```
>>> list1 = ['a','b','c','d']
>>> list1[5]
```

a. ZeroDivisionError

b. ValueError

c. IndexError

d. KeyError

答：c

9. 以下代码将产生什么错误？

```
>>> dict1 = {1:100,2:200,3:300}
>>> dict1[4]
```

答：KeyError

10. 以下代码将产生什么错误？

```
from math import add
```

答：ImportError

11. 以下代码将产生什么错误？

```
>>> a = 1
>>> b ='a'
>>>a + b
```

知识测验，请读者考虑。

12. 查看以下代码。

```
def what_day(day):
    if day in range(1,32):
        print("Valid day")
    else:
        raise ValueError("Invalid day value")
```

以下代码的输出结果是什么？

a. what_day(0)
b. what_day(12)
c. what_day(32)

答：

a. ValueError：Invalid day value
b. Valid day
c. ValueError：Invalid day value

13. 以下代码的输出结果是什么？

```
def just_having_fun():
    try:
            return 45
    except TypeError as e:
            print(str(e))
    finally:
        return("Ok Bye")

jhf = just_having_fun()
print(jhf)
```

答：Ok Bye

第 7 章

数据可视化与数据分析

> **引言** 数据可视化使人脑更容易理解数据模式和趋势，而不考虑数据大小。这些模式隐藏了许多关键信息，对公司非常有益，因为这些隐藏模式分析揭示了需要关注和改进的业务领域。本章介绍如何使用 Python 执行数据可视化和数据分析。

知识结构

- 数据可视化
- Matplotlib
  - Pyplot
  - 绘制点
  - 绘制多点
  - 绘制线
  - 标注 x 轴和 y 轴
- Numpy
  - 安装 Numpy
  - Numpy 数组形状
  - 读取数组元素值
  - 创建 Numpy 数组
- Pandas
- DataFrame 操作

完成本章的学习后，读者应掌握以下知识。
（1）Matplotlib；
（2）Numpy；
（3）Pandas；
（4）Dataframe。

7.1 数据可视化

无论数据集多复杂，良好的可视化有助于赋予其意义。数据可视化并不新鲜，从 Excel 工作表中读取信息对其创建图形（如 Pie 图等）是常见趋势。然而，数据可视化需要更先进、更复杂的技术，如气泡云和树图已经出现。

世界充满竞争，公司在相互竞争中为保持领先地位，需要通过数据可视化掌握运营和整体业务绩效之间的联系。

注：以图形或图表格式表示数据称为数据可视化。数据可视化有助于更快地采取行动，因为图形和图表能够非常有效地汇总复杂数据。无须浏览数千行数据，只需查看图表，大脑就能处理信息并作出正确决定。

下面详细说明数据可视化的优点。

（1）信息以图形或图像形式显示，便于人脑处理。数据可视化有助于理解信息发展的趋势和作出更好的决策。

（2）有助于识别大型数据集的市场变化和趋势。

（3）数据可视化不仅能使人们很好地理解趋势，还可以使人们通过实验对某些变量进行更改以预测变化结果。

（4）确定哪些变量有助于改进流程，哪些变量完全冗余不需要，从而放弃无用信息。

（5）发现漏洞或需要改进之处，并相应带来系统变化。数据可视化有助于识别机会、价值和风险。

（6）有助于查看数据随时间变化的趋势。

（7）提供重要事件随时间变化的信息。

（8）以图形或图表形式生成信息能使人们更好地理解数据集变量之间的相关性。

7.2 Matplotlib

Matplotlib 是 Python 库，类似 MATLAB，用于生成高质量 2D 图形，如直方图、条形图、图表等。Matplotlib 包括多个接口，其中 Pyplot 是本章用到的接口。

Pyplot 包含在 Matplotlib 库中，因此使用前必须安装 Matplotlib 库。安装 Matplotlib 库的步骤如下。

（1）打开图 7.1 所示命令提示符窗口。

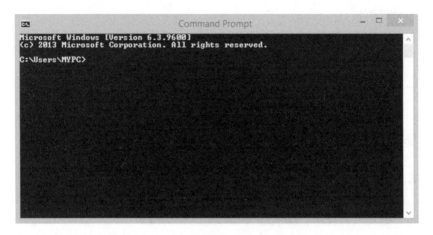

图 7.1

（2）输入图 7.2 所示命令。

图 7.2

（3）Pyplot 安装完成后就可以使用 Matplotlib。

7.2.1 Pyplot

Pyplot 是 Matplotlib 的模块，用来生成 2D 图形框架，常用于 Python 脚本、Shell、Web 应用程序和其他图形用户界面工具包。

在 Python 程序中使用 Pyplot 之前，首先将其导入程序。代码如下所示。

```
import matplotlib.pyplot as plt
```

一般情况下，名称 plt 可以为 –x、y、z、p1 等任意内容，本书使用名称 plt，以便于理解，下面所有示例保留相同的名称。虽然使用不同名称，但工作方式相同。

查看以下语句。

```
import matplotlib.pyplot as plt
```

import matplotlib.pyplot as plt 实际上是从 Matplotlib 库导入 Pyplot 模块，并用名称 plt 将其绑定。

7.2.2 绘制点

下面从绘制点的简单程序开始，如绘制点（5，8）。它需要 4 个简单步骤。

步骤 1：从 Matplotlib 库导入 Pyplot 模块。代码如下所示。

```
import matplotlib.pyplot as plt
```

步骤 2：定义 x = 5 和 y = 8。代码如下所示。

```
x = 5
y = 8
```

步骤 3：调用 plot(x,y) 函数。函数有两个参数：x 代表 x 轴，y 代表 y 轴。代码如下所示。

```
plt.plot(x,y)
```

步骤 4：显示绘图。代码如下所示。

```
plt.show()
```

代码：

```
#导入 Pyplot 模块
import matplotlib.pyplot as plt

#定义 x 和 y 的值
x = 5
y = 8

#调用函数 plot(x,y)画点
plt.plot(x,y)

#显示绘图
plt.show()
```

输出结果如图 7.3 所示。

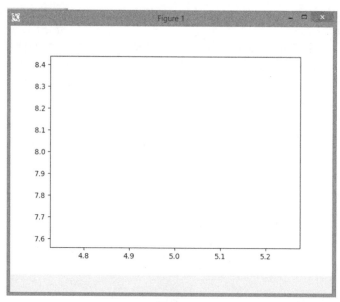

图 7.3

但是点不可见。需要通过定义绘图颜色和样式解决。绘图颜色如表 7.1 所示。

表 7.1

字符	颜色	字符	颜色
'b'	Blue（默认颜色）	'm'	Magenta（洋红）
'g'	Green	'y'	Yellow（黄）
'r'	Red	'k'	Black（黑）
'c'	Cyan	'w'	White（白）

绘图样式如表 7.2 所示。

表 7.2

字符	样式	字符	样式
'-'	solid line style	'o'	circle marker
'--'	dashed line style	'v'	triangle_down marker
'-.'	dash-dot line style	'^'	triangle_up marker
':'	dotted line style	'<'	triangle_left marker
'.'	point marker	'>'	triangle_right marker
','	pixel marker	'1'	tri_down marker

续表

字符	样式	字符	样式
'2'	tri_up marker	'D'	diamond marker
'3'	tri_left marker	'd'	thin_diamond marker
'4'	tri_right marker	'|'	vline marker
's'	square marker	'_'	Hline marker
'O'	Circle	'V'	Down triangle
'p'	pentagon marker	'^'	Up triangle
'*'	star marker	'<'	Left triangle
'h'	hexagon1 marker	'>'	Right triangle
'H'	hexagon2 marker	'S'	Square
'+'	plus marker	'P'	Pentagon
'x'	x marker	'H'	Hexagon

按照表 7.1 和表 7.2 中的信息，将点绘制为红色十字。绘图时将调用函数 plot(x,y,'rx')。

代码如下所示。

```
#导入 Pyplot 模块
import matplotlib.pyplot as plt

#定义 x 和 y 的值
x = 5
y = 8

#调用函数 plot(x,y,'rx')将点绘制为红色十字
plt.plot(x,y,'rx')

#显示绘图
plt.show()
```

现在该点清晰可见，如图 7.4 所示。

跳过步骤 2 直接使用 x 或 y 的值绘制点，这些值可直接传递给 plot 函数，结果一样。

下面的代码跳过步骤 2，直接将值传给函数 plot(x,y)，程序正常运行。

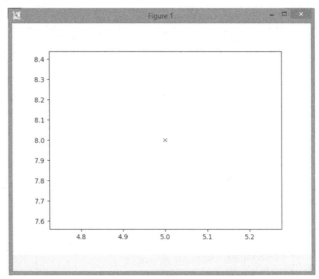

图 7.4

```
#导入 Pyplot 模块
import matplotlib.pyplot as plt

#调用函数 plot(5,8)绘制蓝色六边形点
plt.plot(5,8,'ch')

#显示绘图
plt.show()
```

输出结果如图 7.5 所示。

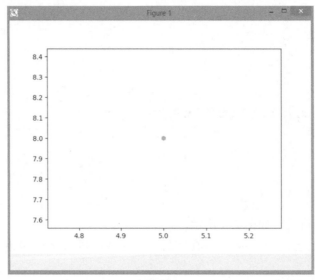

图 7.5

7.2.3 绘制多点

查看表 7.3 所示数据集。

表 7.3

x	10	20	30	40	50	60	70	80	90
y	7	14	21	28	35	42	49	56	63

绘制上述数据集中所有点的步骤如下。

步骤 1：从 Matplotlib 库中导入 Pyplot 模块。代码如下所示。

```
import matplotlib.pyplot as plt
```

步骤 2：定义 x 和 y 列表。代码如下所示。

```
x = [10,20,30,40,50,60,70,80,90]
y = [7,14,21,28,35,42,49,56,63]
```

步骤 3：调用 plot(x,y) 函数。设置点颜色和样式。这里为绿色方块点。代码如下所示。

```
plt.plot(x,y,'gs')
```

步骤 4：显示绘图，如图 7.6 所示。代码如下所示。

```
plt.show()
```

代码：

```
#从 Matplotlib 库导入 Pyplot 模块
import matplotlib.pyplot as plt

#定义列表 x 和列表 y 的值
x = [10,20,30,40,50,60,70,80,90]
y = [7,14,21,28,35,42,49,56,63]

#调用函数 plot(x,y),绘制绿色方块图
plt.plot(x,y,'gs')

#显示绘图
plt.show()
```

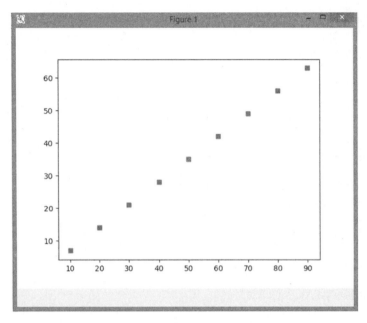

图 7.6

7.2.4 绘制线

绘制线类似绘制多点。跳过设置点的颜色与标注样式。使用与上个示例相同的数据集，如表 7.3 所示。

绘制线的步骤如下。

步骤 1：从 Matplotlib 库导入 Pyplot 模块。代码如下所示。

```
import matplotlib.pyplot as plt
```

步骤 2：定义 x 和 y 列表。代码如下所示。

```
x = [10,20,30,40,50,60,70,80,90]
y = [7,14,21,28,35,42,49,56,63]
```

步骤 3：调用 plot(x,y) 函数（不设置颜色与标注样式）。代码如下所示。

```
plt.plot(x,y)
```

步骤 4：显示绘图。代码如下所示。

```
plt.show()
```

代码:

```
#从 Matplotlib 库导入 Pyplot 模块
import matplotlib.pyplot as plt

#定义列表 x 和 y 的值
x = [10,20,30,40,50,60,70,80,90]
y = [7,14,21,28,35,42,49,56,63]

#调用函数 plot(x,y)绘制线
plt.plot(x,y)

#显示绘图
plt.show()
```

输出结果如图 7.7 所示。

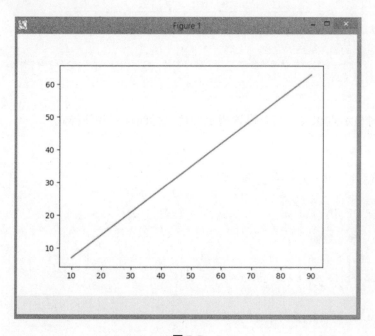

图 7.7

线标注符不同于点标注符。线标注符如下。

(1) '-': 实线;

(2) '--': 虚线;

(3) ":": 点线;

(4) '-.': 点划线。

对前面的示例稍作修改,绘制洋红色点线。

输出结果如图 7.8 所示。

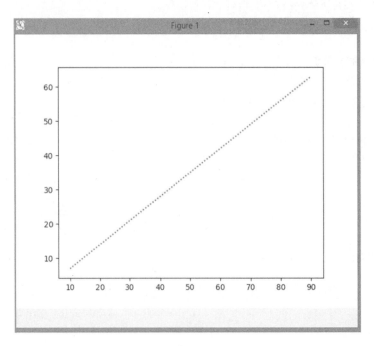

图 7.8

如果同时显示点和线,需要设置指定颜色、标注样式和线样式。

代码:

```
#从 Matplotlib 库导入 Pyplot 模块
import matplotlib.pyplot as plt

#定义列表 x 和 y 的值
x = [10,20,30,40,50,60,70,80,90]
y = [7,14,21,28,35,42,49,56,63]

#调用 plot(x,y)函数,绘制洋红色点线
plt.plot(x,y,"mo:")

#显示绘图
plt.show()
```

输出结果如图 7.9 所示。

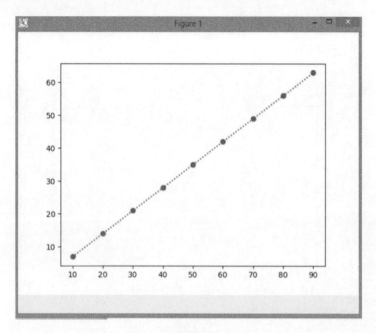

图 7.9

另外，可以设置点的大小（标注大小）和线的宽度（线宽）。
代码：

```
#从 Matplotlib 库导入 Pyplot 模块
import matplotlib.pyplot as plt

#定义列表 x 和 y 的值
x = [10,20,30,40,50,60,70,80,90]
y = [7,14,21,28,35,42,49,56,63]

#调用函数 plot(x,y),绘制指定大小的点和指定宽度的线
plt.plot(x,y,"mo:", markersize = 9, linewidth = 3)

#显示绘图
plt.show()
```

输出结果如图 7.10 所示。

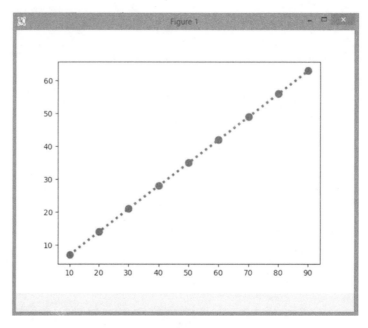

图 7.10

如果背景显示网格，则使用 grid( ) 函数。
代码：

```
#从 Matplotlib 库导入 Pyplot 模块
import matplotlib.pyplot as plt

#定义列表 x 和 y 的值
x = [10,20,30,40,50,60,70,80,90]
y = [7,14,21,28,35,42,49,56,63]

#调用函数 plot(x,y) 绘制洋红色点线图
plt.plot(x,y,"mo:", markersize = 9, linewidth = 3)

#显示网格
plt.grid()

#显示绘图
plt.show()
```

输出结果如图 7.11 所示。

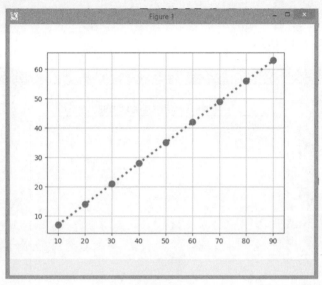

图 7.11

7.2.5 标注 x 轴和 y 轴

函数 xlabel( ) 和 ylabel( ) 用于标注 x 轴和 y 轴，函数 title( ) 用于设置图的标题。示例实现过程如下。

代码：

```
#从 Matplotlib 库导入 Pyplot 模块
import matplotlib.pyplot as plt

#定义列表 x 和列表 y 的值
x = [10,20,30,40,50,60,70,80,90]
y = [7,14,21,28,35,42,49,56,63]

#调用函数 plot(x,y)
plt.plot(x,y)

#标注 x 轴为 X-axis
plt.xlabel('X-axis')

#标注 y 轴为 Y-axis
plt.ylabel('Y-axis')

#调用函数 title()设置图的标题
plt.title('Simple Line Plot using PyPlot')
```

```
#显示绘图
plt.show()
```

输出结果如图 7.12 所示。

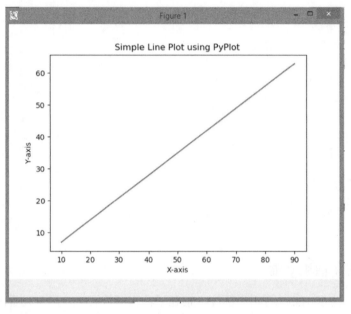

图 7.12

下一个程序在同一图上绘制两条直线,需要两个数据集,如表 7.4 和表 7.5 所示。

表 7.4

x1	10	20	30	40	50	60	70	80	90
y1	7	14	21	28	35	42	49	56	63

表 7.5

x2	7	14	21	28	35	42	49	56	63
y2	10	20	30	40	50	60	70	80	90

步骤如下。

步骤 1:从 Matplotlib 库导入 Pyplot 模块。代码如下所示。

```
import matplotlib.pyplot as plt
```

步骤 2:定义列表 x1 和 y1。代码如下所示。

```
x1 = [10,20,30,40,50,60,70,80,90]
y1 = [7,14,21,28,35,42,49,56,63]
```

步骤3：调用函数 plot(x,y) 读取列表 x1 和 y1。代码如下所示。

```
plt.plot(x1,y1)
```

步骤4：定义 x2 和 y2 列表。代码如下所示。

```
x2 = [7,14,21,28,35,42,49,56,63]
y2 = [10,20,30,40,50,60,70,80,90]
```

步骤5：调用函数 plot(x,y) 计算 x2 和 y2。代码如下所示。

```
plt.plot(x2,y2)
```

步骤6：标注 x 轴。代码如下所示。

```
plt.xlabel('X-axis')
```

步骤7：标注 y 轴。代码如下所示。

```
plt.ylabel('Y-axis')
```

步骤8：设置图标题。代码如下所示。

```
plt.title('Plotting two lines on the same graph using PyPlot')
```

步骤9：显示绘图。代码如下所示。

```
plt.show()
```

代码：

```
#从 Matplotlib 库导入 Pyplot 模块
import matplotlib.pyplot as plt

#定义列表 x1 和 y1 的值
x1 = [10,20,30,40,50,60,70,80,90]
y1 = [7,14,21,28,35,42,49,56,63]

#调用函数 plot(x,y) 计算列表 x1 和 y1
plt.plot(x1,y1)

#定义列表 x2 和 y2 的值
x2 = [7,14,21,28,35,42,49,56,63]
y2 = [10,20,30,40,50,60,70,80,90]
```

```
#调用函数 plot(x,y)计算列表 x2 和 y2 的值
plt.plot(x2,y2)

#标注 x 轴
plt.xlabel('X - axis')

#标注 y 轴
plt.ylabel('Y - axis')

#设置图标题
plt.title('Plotting two lines on the same graph using PyPlot')

#显示绘图
plt.show()
```

输出结果如图 7.13 所示。

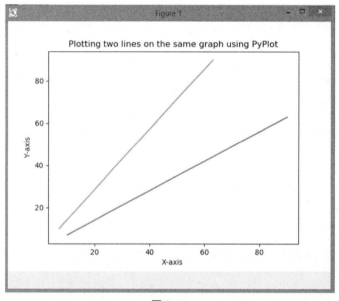

图 7.13

从图 7.13 中可以发现，图标正常，但无法确定哪条线对应哪个数据集。使用函数 legend()标注线与数据集的对应关系。

legend()函数用于创建小矩形框，表示哪条线代表哪个数据集信息。数据集在调用绘图函数时为该图定义标签。在调用函数 legend()时，线类型与数据集标签同时显示。

代码：

```
#从 Matplotlib 库导入 Pyplot 模块
import matplotlib.pyplot as plt
```

```
#定义列表 x1 和 y1
x1 = [10,20,30,40,50,60,70,80,90]
y1 = [7,14,21,28,35,42,49,56,63]

#调用函数 plot(x,y)计算列表 x1 和 y1 并设置图标签
plt.plot(x1,y1, label ='Line for dataset x1, y1')

#定义列表 x2 和 y2
x2 = [7,14,21,28,35,42,49,56,63]
y2 = [10,20,30,40,50,60,70,80,90]

#调用函数 plot(x,y)计算列表 x2 和 y2 并设置图标签
plt.plot(x2,y2, label ='Line for dataset x2, y2')

#标注 x 轴
plt.xlabel('X-axis')

#标注 y 轴
plt.ylabel('Y-axis')

#设置图标题
plt.title('Plotting two lines on the same graph using PyPlot')

#调用函数 legend()
plt.legend()

#显示绘图
plt.show()
```

输出结果如图 7.14 所示。

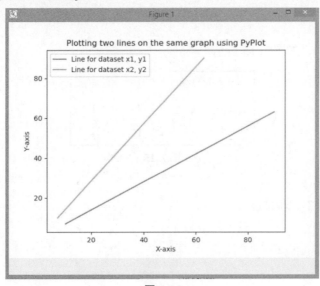

图 7.14

蓝线显示 x1 和 y1 曲线图，红线显示 x2 和 y2 曲线图。

7.3 Numpy

本节介绍 Numpy 数组。从安装 Numpy 模块开始本节内容。

7.3.1 安装 Numpy

安装 Numpy 的步骤如下。
（1）打开命令提示符窗口。
（2）将目录改为 Python 安装目录（与安装 Matplotlib 的目录一致）。
（3）输入命令 pip install Numpy。

7.3.2 Numpy 数组形状

Numpy 数组有维度，通过关键字 shape 得到。数组形状是每个维度元素个数的元组。代码如下所示。

```
>>> import numpy as np
>>> arr = np.array([[1,2,3,4],
      [3,5,2,1],
      [6,4,1,4]])
```

打印数组 arr：

```
[[1 2 3 4]
 [3 5 2 1]
 [6 4 1 4]]
```

输出 3 行 4 列，如图 7.15 所示。

1	2	3	4
3	5	2	1
6	4	1	4

图 7.15

该数组形状是 3×4。
使用 shape 命令打印维度。
代码：

```
print(arr.shape)
```

输出：

```
(3,4)
```

7.3.3 读取数组元素值

使用索引读取数组元素值。

输入以下命令:

```
print(arr[2][0])
```

输出结果为 6(见图 7.16)。

	0	1	2	3
0	1	2	3	4
1	3	5	2	1
2	6	4	1	4

图 7.16

arr 对应索引 2 处元素为 [6, 4, 1, 4]。

arr 对应索引 2 处元素的索引 0 处元素为 6。

同样,对于以下代码:

```
print(arr[0][0])
```

输出结果如图 7.17 所示。

	0	1	2	3
0	1	2	3	4
1	3	5	2	1
2	6	4	1	4

图 7.17

7.3.4 创建 Numpy 数组

创建 Numpy 数组。代码如下所示。

```
numpy.linspace()
```

linspace()方法用于返回指定间隔内的均匀间隔数字。

语法格式如下所示。

```
numpy.linspace(start,stop,num)
```

(1) start:强制值,设置序列起点。

(2) stop:强制值,设置序列终点。

(3) num:可选值,设置在开始值和停止值之间生成值的数量。

代码：

```
import numpy as np
arr = np.linspace(3,5)
print(arr)
```

输出结果如图 7.18 所示。

[3.	3.04081633	3.08163265	3.12244898	3.16326531	3.20408163
3.24489796	3.28571429	3.32653061	3.36734694	3.40816327	3.44897959
3.48979592	3.53061224	3.57142857	3.6122449	3.65306122	3.69387755
3.73469388	3.7755102	3.81632653	3.85714286	3.89795918	3.93877551
3.97959184	4.02040816	4.06122449	4.10204082	4.14285714	4.18367347
4.2244898	4.26530612	4.30612245	4.34693878	4.3877551	4.42857143
4.46938776	4.51020408	4.55102041	4.59183673	4.63265306	4.67346939
4.71428571	4.75510204	4.79591837	4.83673469	4.87755102	4.91836735
4.95918367	5.	]			

图 7.18

这是因为没有指定开始值和停止值之间生成值的数量。

假设生成 2 和 3 之间的 10 个值。

代码：

```
import numpy as np
arr = np.linspace(2,3,10)
print(arr)
```

输出结果如图 7.19 所示。

[2.	2.11111111	2.22222222	2.33333333	2.44444444
2.55555556				
2.66666667	2.77777778	2.88888889	3.	]

图 7.19

下面介绍 Numpy.arange() 函数。

Numpy.arange() 函数的语法格式如下所示。

```
Numpy.arange(start,stop,step)
```

（1）start：序列起点。

（2）stop：序列终点。

（3）step：两连续值之间的间隔，默认设置为 1。

示例 7.1

写出以下代码的输出结果。

```
>>> arr = np.arange(2,3,10)
>>> print(arr)
```

答:[2]。

示例 7.2

写出以下代码的输出结果。

```
arr = np.arange(1,9,2)
print(arr)
```

答:[1 3 5 7]。

示例 7.3

写出以下代码的输出结果。

```
arr = np.arange(2,5)
print(arr)
```

答:[2 3 4]。

查看使用 x 和 y 的示例。如果变量之间有依赖关系,如以 y = mx + c 绘制直线。有 x 值,但需要计算 y 值。Numpy 也可以实现这一点。

为 x 值绘制图形,如表 7.6 所示。

表 7.6

| x | 1 | 4 | 7 | 9 | 10 | 15 |

直线公式为

$$y = 4x + 6$$

代码:

```
import numpy as np
import matplotlib.pyplot as plt

x = np.array([1,4,7,9,10,15])
y = 4 * x + 6
plt.plot(x,y,"rs--", markersize = 6, linewidth = 3)
plt.xlabel("x")
plt.ylabel("y = 4x + 6")
plt.show()
```

输出结果如图 7.20 所示。

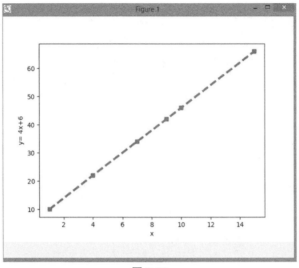

图 7.20

示例 7.4

绘制 f(x) 与 x 的关系图,其中 x 在 20~40 范围内,间隔 5 个值,f(x) = log(x)。绘图步骤如下。

步骤 1:导入 Numpy。代码如下所示。

```
import numpy as np
```

步骤 2:导入 Pyplot。代码如下所示。

```
import matplotlib.pyplot as plt
```

步骤 3:定义数组从 20 到 40,相隔 5 个值。代码如下所示。

```
x = np.arange(20,40,5)
```

步骤 4:定义 f(x) = log(x)。代码如下所示。

```
y = np.log(x)
```

步骤 5:调用 plot() 函数。代码如下所示。

```
plt.plot(x,y)
```

步骤 6:标注 x 轴。代码如下所示。

```
plt.xlabel("x values from 20 -40, 5 steps apart")
```

步骤 7:标注 y 轴。代码如下所示。

```
plt.ylabel("f(x) = log(x)")
```

步骤 8:设置标题。代码如下所示。

```
plt.title("f(x) vs. x")
```

步骤9：显示绘图。代码如下所示。

```
plt.show()
```

代码：

```
#导入 Numpy
import numpy as np

#导入 Pyplot
import matplotlib.pyplot as plt

#定义数组从 20 到 40,相隔 5 个值
x = np.arange(20,40,5)

#定义 f(x) = log(x)
y = np.log(x)

#调用 plot() 函数
plt.plot(x,y)

#标注 x 轴
plt.xlabel("x values from 20 -40, 5 steps apart")

#标注 y 轴
plt.ylabel("f(x) = log(x)")

#设置标题
plt.title("f(x) vs. x")

#显示绘图
plt.show()
```

输出结果如图 7.21 所示。

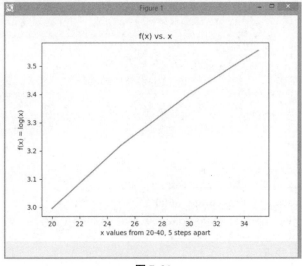

图 7.21

Numpy 主要用于处理多维数组。

示例7.5

从用户处读取一个合适列表作为输入,并绘制 y 的图形。

代码:

```python
#导入 Pyplot
import matplotlib.pyplot as plt

#导入 Numpy
import numpy as np

#确定 start、end 和 step 的值
start = int(input("Enter the starting point of the range : "))
end = int(input("Enter the end point of the range : "))
step = int(input("Enter the interval between two consecutive values : "))

#定义数组 x
x = np.arange(start,end,step)

#定义数组 y
y = (x**2)+(3*x)

#调用 plot(x,y)函数
plt.plot(x,y)

#标注 x 轴
plt.xlabel("X-axis")

#标注 y 轴
plt.ylabel("Y-axis")

#绘制标题
plt.title("plotting of function")

#显示绘图
plt.show()
```

输出:

```
Enter the starting point of the range : 100
Enter the end point of the range : 600
Enter the interval between two consecutive values : 30
```

输出结果如图7.22所示。

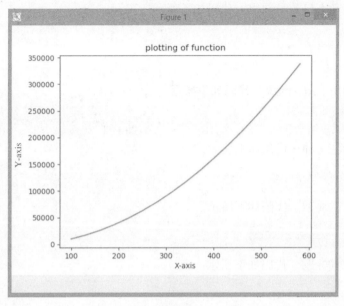

图 7.22

下面介绍如何绘制 Bar 图。

示例 7.6

以 Bar 图表示表 7.7 所提供的国家及其人口数量。

表 7.7

Country	Population
China	1384688986
India	1296834042
USA	329256465
Indonesia	262787403
Brazil	208846892
Nigeria	195300343
Bangladesh	159453001

步骤 1：导入 Pyplot。代码如下所示。

```
import matplotlib.pyplot as plt
```

步骤 2：定义国家名称数组。代码如下所示。

```
x = ['China','India','USA','Indonesia','Brazil','Nigeria','Bangladesh']
```

步骤3：定义人口数量数组。代码如下所示。

```
y = [1384688986, 1296834042, 329256465, 262787403, 208846892, 195300343, 159453001]
```

步骤4：调用bar()函数。代码如下所示。

```
plt.bar(x,y)
```

步骤5：标注x轴。代码如下所示。

```
plt.xlabel("Country")
```

步骤6：标注y轴。代码如下所示。

```
plt.ylabel("Population")
```

步骤7：设置标题。代码如下所示。

```
plt.title("Country vs. Population")
```

步骤8：显示绘图。代码如下所示。

```
plt.show()
```

代码：

```
#导入Pyplot
import matplotlib.pyplot as plt

#定义国家名称数组
x = ['China','India','USA','Indonesia','Brazil','Nigeria','Bangladesh']

#定义人口数量数组
y = [1384688986, 1296834042, 329256465, 262787403, 208846892, 195300343, 159453001]

#调用bar()函数
plt.bar(x,y)

#标注x轴
plt.xlabel("Country")

#标注y轴
plt.ylabel("Population")
```

```
#设置标题
plt.title("Country vs. Population")

#显示绘图
plt.show()
```

输出结果如图 7.23 所示。

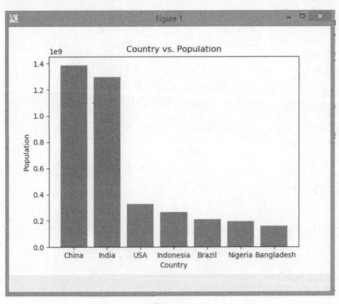

图 7.23

示例 7.7

使用 Python 编写程序，x 轴显示学生编号（student id），y 轴以 Bar 图形式显示两门课（Science 和 Commerce）的百分比分数（Score），如表 7.8 所示。

表 7.8

Science		Commerce	
Student id	Score/%	Student id	Score/%
101	78	102	89
104	57	103	100
108	89	105	98
109	99	106	67
110	86	107	100
112	72	111	20

代码：

```python
#导入 Pyplot
import matplotlib.pyplot as plt

#定义 Science 学科学生的 Student id 和 Score 数组
x1 = [101,104,108,109,110,112]
y1 = [78,57,89,99,86,72]

#定义 Commerce 学科学生的 Student id 和 Score 数组
x2 = [102,103,105,106,107,111]
y2 = [89,100,98,67,100,20]

#调用 plot()函数
plt.bar(x1,y1,label='Science department')
plt.bar(x2,y2,label='Commerce department')

#标注 x 轴
plt.xlabel("Student id")

#标注 y 轴
plt.ylabel("%Score")

#设置标题
plt.title("Score for Science and Commerce Students")

plt.legend()

#显示绘图
plt.show()
```

输出结果如图 7.24 所示。

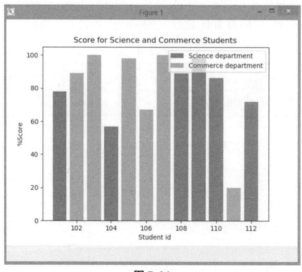

图 7.24

示例7.8

表7.9显示学生的考试总分（满分为500），并将分数转换为百分比，以 Bar 图显示。

表7.9

Student name	Score（满分为500）
John	376
Ted	455
Lizy	489
Elena	300
Alex	250
Albert	340
Meg	400

代码：

```python
#导入 Pyplot
import matplotlib.pyplot as plt

#导入 Numpy
import numpy as np

#定义 Student name 数组
x = ['John','Ted','Lizy','Elena','Alex','Albert','Meg']

#定义 Score 数组
arr = np.array([376,455,489,300,250,340,400])
y = arr*100/500

#调用 bar()函数
plt.bar(x,y)

#标注 x 轴
plt.xlabel("Student name")

#标注 y 轴
plt.ylabel("%Score")

#设置标题
plt.title("Score")

#显示绘图
plt.show()
```

输出结果如图 7.25 所示。

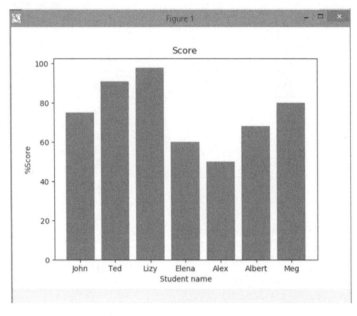

图 7.25

下面介绍如何绘制 Pie 图。

示例 7.9

表 7.10 表示空气成分（Gas）的百分比（Percentage）。将此数据以 Pie 图表示。

表 7.10

Gas	Percentage/%
Nitrogen	77.8
O_2	19.9
CO_2	0.3
others	2.0

代码：

```
import matplotlib.pyplot as plt
# 数据
gases = ['Nitrogen','O₂','CO₂','others']
sizes = [78,20,0.3,1.97]
colors = ['pink','yellow','orange','lightskyblue']
```

```
#绘图
plt.pie(sizes,explode=(0,0.1,0,0),labels=gases,colors=colors,
autopct='%1.1f%%', shadow=True, startangle=140)
plt.legend()
plt.show()
```

输出结果如图 7.26 所示。

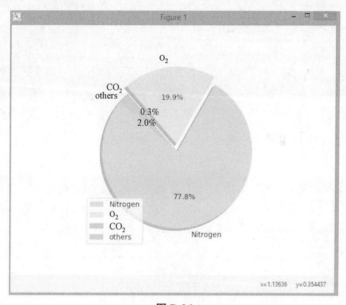

图 7.26

7.4 Pandas

本节将学习 Pandas。Pandas 是一款快速、易于使用且功能强大的数据分析/操作工具，以灵活性著称。它是 Python 程序员处理数据的首选库，如图 7.27 所示。

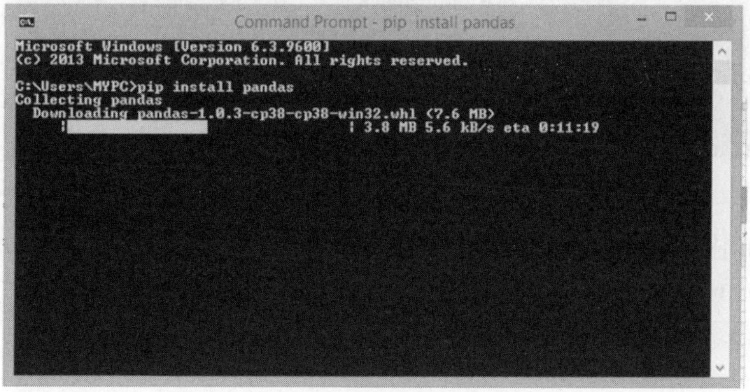

图 7.27

Pandas 安装成功后，将显示在命令提示符下，如图 7.28 所示。

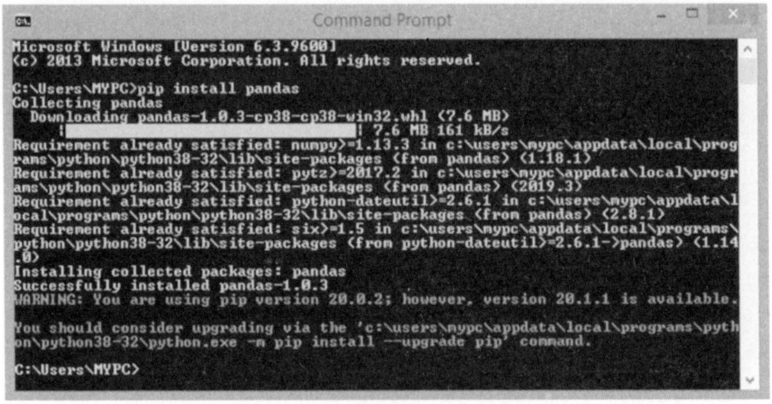

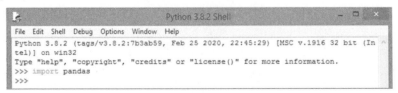

图 7.28

在 Python Shell 中检查是否已安装 Pandas。如果安装正确，则不会产生错误，如图 7.29 所示。

图 7.29

如果要使用 Pandas，还需要使用 pip 安装 xlrd 包，如图 7.30 所示。

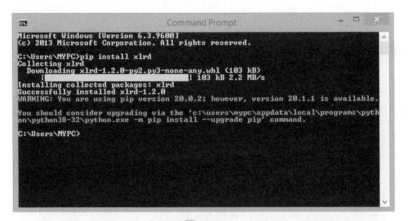

图 7.30

下面使用 Excel 工作表。对于这个示例，要使用 "F:\data" 目录中的 "shares.xlsx" 文件，如图 7.31 所示。

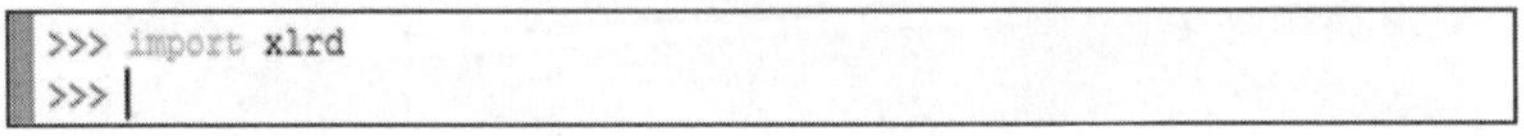

```
>>> import xlrd
>>>
```

图 7.31

1. 基于 Excel 工作表创建 DataFrame

首先学习如何使用 Pandas 访问 Excel 文件。示例中访问"F:\data"目录下的"shares.xlsx"文件，数据在 Excel 的工作表 1 中，如图 7.32 所示。

	A	B	C	D	E	F	G
1	share	quantity	price	description	amount invested	profit%	
2	Coal india	12	311.7	POWER	3740.4	15%	
3	Powergrid	16	97.88		1566.08	7%	
4	Polyplex Corporatic	10	484	4th largest manufacturer of thin polyester Fil	4840		
5	noida toll	40	33.5		1340		
6	Kotak bank	5	479.85		2399.25	2.00%	
7	Sail	10	187.75		1877.5	4%	
8	SBI	3	2756.6		8269.8		
9	SBI	2	2191.5		4383		
10	SBI	2	1911.5		3,823		
11	Reliance	5	773.5		3867.5	11%	
12	JP Associate	13	59.5		773.5	19%	
13	suzlon	13	38.78		504.14		
14	tata steel	10	415.87		4158.7	14%	
15	SAIL	10	109.7		1097		
16	ICICI	3	857.3		2571.9	8%	
17	IDFC	10	110.45		1104.5	12%	
18						15%	
19	SAIL	9	104.1		936.9		
20	IDFC	9	108		972	15%	
21	SUZLON	13	37.2		483.6		
22	tatamotors	5	154.85		774.25	21%	
23	ICICI	3	780.7		2342.1	18%	
24	icici	6	786.25				
25	sbi	2	1727.55				
26	icici	1	722.1				
27	sbi	1	1636.35				
28	icici	1	713.4				
29	Total				51825.12		
30							

图 7.32

步骤 1：导入 Pandas。代码如下所示。

```
>>> import pandas as pd
```

步骤 2：打开 Excel 文件读取内容。代码如下所示。

```
>>> df = pd.read_excel("F:\\data\shares.xlsx",'Sheet1')
>>> df
```

其中，df 是 DataFrame 对象。DataFrame 是二维数据结构，大小可变，表中数据以行和列的表格格式存储。

代码：

```
>>> import pandas as pd
>>> df = pd.read_excel("F:\\data\shares.xlsx",'Sheet1')
>>> df
```

输出结果如图 7.33 所示。

```
>>> df = pd.read_excel("F:\\data\shares.xlsx",'Sheet1')
>>> df
              share  quantity  ...  amount invested  profit%
0          Coal india     12.0  ...          3740.40     0.15
1           Powergrid     16.0  ...          1566.08     0.07
2  Polyplex Corporation  10.0  ...          4840.00      NaN
3          noida toll    40.0  ...          1340.00      NaN
4          Kotak bank     5.0  ...          2399.25     0.02
5                Sail    10.0  ...          1877.50     0.04
6                 SBI     3.0  ...          8269.80      NaN
7                 SBI     2.0  ...          4383.00      NaN
8                 SBI     2.0  ...          3823.00      NaN
9             Reliance     5.0  ...          3867.50     0.11
10        JP Associate   13.0  ...           773.50     0.19
11              suzlon   13.0  ...           504.14      NaN
12          tata steel   10.0  ...          4158.70     0.14
13                SAIL   10.0  ...          1097.00      NaN
14               ICICI    3.0  ...          2571.90     0.08
15                IDFC   10.0  ...          1104.50     0.12
16                 NaN    NaN  ...              NaN     0.15
17                SAIL    9.0  ...           936.90      NaN
18                IDFC    9.0  ...           972.00     0.15
19              SUZLON   13.0  ...           483.60      NaN
20          tatamotors    5.0  ...           774.25     0.21
21               ICICI    3.0  ...          2342.10     0.18
22               icici    6.0  ...              NaN      NaN
23                 sbi    2.0  ...              NaN      NaN
24               icici    1.0  ...              NaN      NaN
25                 sbi    1.0  ...              NaN      NaN
26               icici    1.0  ...              NaN      NaN
27               Total    NaN  ...         51825.12      NaN

[28 rows x 6 columns]
```

图 7.33

注意，DataFrame 对象在左侧显示一个附加列。

2. 基于 ".csv" 文件创建 DataFrame

以 ".csv" 格式存储相同文件。基于 ".csv" 文件创建 DataFrame 的步骤如下。

步骤 1：导入 Pandas。代码如下所示。

```
import pandas as pd
```

步骤 2：读取 ".csv" 文件并将数据存储在 DataFrame 对象中。代码如下所示。

```
>>> df = pd.read_csv("F:\\data\shares.csv")
>>> df
```

输出结果如图 7.34 所示。

```
        share        quantity  ...  amount invested  profit%
0       Coal india   12.0      ...  3740.4           15%
1       Powergrid    16.0      ...  1566.08          7%
2       Polyplex Corporation  10.0  ...  4840        NaN
3       noida toll   40.0      ...  1340             NaN
4       Kotak bank   5.0       ...  2399.25          2.00%
5       Sail         10.0      ...  1877.5           4%
6       SBI          3.0       ...  8269.8           NaN
7       SBI          2.0       ...  4383             NaN
8       SBI          2.0       ...  3,823            NaN
9       Reliance     5.0       ...  3867.5           11%
10      JP Associate 13.0      ...  773.5            19%
11      suzlon       13.0      ...  504.14           NaN
12      tata steel   10.0      ...  4158.7           14%
13      SAIL         10.0      ...  1097             NaN
14      ICICI        3.0       ...  2571.9           8%
15      IDFC         10.0      ...  1104.5           12%
16      NaN          NaN       ...  NaN              15%
17      SAIL         9.0       ...  936.9            NaN
18      IDFC         9.0       ...  972              15%
19      SUZLON       13.0      ...  483.6            NaN
20      tatamotors   5.0       ...  774.25           21%
21      ICICI        3.0       ...  2342.1           18%
22      icici        6.0       ...  NaN              NaN
23      sbi          2.0       ...  NaN              NaN
24      icici        1.0       ...  NaN              NaN
25      sbi          1.0       ...  NaN              NaN
26      icici        1.0       ...  NaN              NaN
27      Total        NaN       ...  51825.12         NaN
```

图 7.34

3. 基于字典创建 DataFrame

字典数据存储在键 – 值对中。基于字典创建 DataFrame 的步骤如下。

步骤 1：导入 Pandas。代码如下所示。

```
import pandas as pd
```

步骤 2：定义字典对象。代码如下所示。

```
shares = {"share":["Coal india","Powergrid","Polyplex Corporation","noida toll","kotakbank","Sail"],"quantity":[12.0,16.0,10.0,40.0,5.0,10.0],"amount invested":[3740.4,1566.08,4840.0,1340.0,2399.25,1877.5],"profit":[15,7,0,0,2,4]}
```

步骤 3：创建 DataFrame 对象。代码如下所示。

```
df = pd.DataFrame(shares)
```

代码：

```
>>> import pandas as pd
>>> shares = {"share":["Coal india","Powergrid","Polyplex Corporation",
"noida toll","kotakbank","Sail"],"quantity":[12.0,16.0,10.0,40.0,5.0,
10.0],"amount invested":[3740.4,1566.08,4840.0,1340.0,2399.25,1877.5]," -
profit":[15,7,0,0,2,4]}
>>> df = pd.DataFrame(shares)
>>> df
```

输出结果如图 7.35 所示。

```
                   share  quantity  amount invested  profit
0             Coal india      12.0          3740.40      15
1              Powergrid      16.0          1566.08       7
2   Polyplex Corporation      10.0          4840.00       0
3             noida toll      40.0          1340.00       0
4            kotak bank       5.0          2399.25       2
5                   Sail      10.0          1877.50       4
>>>
```

图 7.35

4. 基于元组列表创建 DataFrame

上面基于字典创建了 DataFrame。下面介绍如何基于元组列表创建 DataFrame。

步骤 1：导入 Pandas。代码如下所示。

```
import pandas as pd
```

步骤 2：定义元组列表。代码如下所示。

```
Shares = [("Coal india",12.0,3740.4,15),("Powergrid",16.0, 1566.08,7),
("Polyplex Corporation",10.0,4840.0,0),("noida toll",40.0, 1340.0,0),
("kotak bank",5.0,2399.25,2),("Sail",10.0,1877.5,4)]
```

步骤 3：创建 DataFrame 对象。

传递元组列表，并定义列名。代码如下所示。

```
df = pd.DataFrame(shares,columns =["share","quantity","amountinvested",
"profit"])
```

代码：

```
import pandas as pd
shares = [("Coal india",12.0,3740.4,15),("Powergrid",16.0, 1566.08,7),
("Polyplex Corporation",10.0,4840.0,0),("noida toll",40.0, 1340.0,0),
("kotak bank",5.0,2399.25,2),("Sail",10.0,1877.5,4)]
df = pd.DataFrame(shares,columns =["share","quantity","amountinvested",
"profit"])
print(df)
```

输出结果如图 7.36 所示。

```
          share       quantity   amount invested   profit
0         Coal india      12.0             3740.40     15
1         Powergrid       16.0             1566.08      7
2  Polyplex Corporation   10.0             4840.00      0
3         noida toll      40.0             1340.00      0
4         kotak bank       5.0             2399.25      2
5               Sail      10.0             1877.50      4
>>>
```

图 7.36

7.5 DataFrame 操作

本节介绍如何获取 DataFrame 的行数和列数。

基于 Excel 工作表创建 DataFrame。代码如下所示。

```
>>> import pandas as pd
>>> df = pd.read_excel("F:\\data\shares.xlsx","Sheet1")
```

使用 shape 属性可以获取 DataFrame 对象的行数和列数。代码如下所示。

```
>>> df.shape
(28,6)
```

为了清楚起见,代码可以这样写:

```
>>> row,col = df.shape
>>> print("The dataframe has {} rows.".format(row))
The dataframe has 28 rows.
>>> print("The dataframe has {} columns.".format(col))
The dataframe has 6 columns.
```

1. 检索行和列数据

使用 head() 函数查看 DataFrame 的前 5 行。

代码:

```
>>> import pandas as pd
>>> df = pd.read_excel("F:\\data\shares.xlsx","Sheet1")
>>> df.head()
```

输出结果如图 7.37 所示。

```
          share       quantity  ...   amount invested  profit%
0         Coal india      12.0  ...          3740.40     0.15
1         Powergrid       16.0  ...          1566.08     0.07
2  Polyplex Corporation   10.0  ...          4840.00      NaN
3         noida toll      40.0  ...          1340.00      NaN
4         Kotak bank       5.0  ...          2399.25     0.02
```

图 7.37

同样，使用 tail( )函数查看 DataFrame 的最后 5 行。
代码：

```
>>> df.tail()
```

输出结果如图 7.38 所示。

```
>>> df.tail()
      share  quantity    price  description  amount invested  profit%
23      sbi       2.0  1727.55          NaN              NaN      NaN
24    icici       1.0   722.10          NaN              NaN      NaN
25      sbi       1.0  1636.35          NaN              NaN      NaN
26    icici       1.0   713.40          NaN              NaN      NaN
27    Total       NaN      NaN          NaN         51825.12      NaN
```

图 7.38

设置 head( )或 tail( )函数的参数检索少于前 5 行或 5 列，如图 7.39 所示。

```
>>> df.head(3)
                  share  quantity  ...  amount invested  profit%
0            Coal india      12.0  ...          3740.40     0.15
1             Powergrid      16.0  ...          1566.08     0.07
2   Polyplex Corporation     10.0  ...          4840.00      NaN

[3 rows x 6 columns]
>>> df.tail(1)
     share  quantity  price  description  amount invested  profit%
27   Total       NaN    NaN          NaN         51825.12      NaN
>>>
```

图 7.39

使用切片检索列行，用法和列表一样，如图 7.40 所示。

```
>>> df[4:16]
          share  quantity    price  description  amount invested  profit%
4    Kotak bank       5.0   479.85          NaN          2399.25     0.02
5          Sail      10.0   187.75          NaN          1877.50     0.04
6           SBI       3.0  2756.60          NaN          8269.80      NaN
7           SBI       2.0  2191.50          NaN          4383.00      NaN
8           SBI       2.0  1911.50          NaN          3823.00      NaN
9      Reliance       5.0   773.50          NaN          3867.50     0.11
10  JP Associate      13.0    59.50          NaN           773.50     0.19
11       suzlon      13.0    38.78          NaN           504.14      NaN
12    tata steel      10.0   415.87          NaN          4158.70     0.14
13          SAIL      10.0   109.70          NaN          1097.00      NaN
14         ICICI       3.0   857.30          NaN          2571.90     0.08
15          IDFC      10.0   110.45          NaN          1104.50     0.12
```

图 7.40

也可以使用类似列表或字符串的字符串运算符以逆序显示行。

代码：

```
>>> df[::-1]
```

输出结果如图 7.41 所示。

```
>>> df[::-1]
        share         quantity  ...  amount invested  profit%
27      Total              NaN  ...         51825.12      NaN
26      icici             1.0   ...              NaN      NaN
25        sbi             1.0   ...              NaN      NaN
24      icici             1.0   ...              NaN      NaN
23        sbi             2.0   ...              NaN      NaN
22      icici             6.0   ...              NaN      NaN
21       ICICI            3.0   ...          2342.10     0.18
20   tatamotors           5.0   ...           774.25     0.21
19       SUZLON          13.0   ...           483.60      NaN
18         IDFC           9.0   ...           972.00     0.15
17         SAIL           9.0   ...           936.90      NaN
16          NaN           NaN  ...              NaN     0.15
15         IDFC          10.0   ...          1104.50     0.12
14        ICICI           3.0   ...          2571.90     0.08
13         SAIL          10.0   ...          1097.00      NaN
12    tata steel         10.0   ...          4158.70     0.14
11        suzlon         13.0   ...           504.14      NaN
10   JP Associate        13.0   ...           773.50     0.19
9        Reliance         5.0   ...          3867.50     0.11
8             SBI         2.0   ...          3823.00      NaN
7             SBI         2.0   ...          4383.00      NaN
6             SBI         3.0   ...          8269.80      NaN
5            Sail        10.0   ...          1877.50     0.04
4       Kotak bank        5.0   ...          2399.25     0.02
3       noida toll       40.0   ...          1340.00      NaN
2   Polyplex Corporation 10.0   ...          4840.00      NaN
1         Powergrid      16.0   ...          1566.08     0.07
0        Coal india      12.0   ...          3740.40     0.15
[28 rows x 6 columns]
```

图 7.41

2. 检索指定列

使用 column 属性可以检索 DataFrame 的列名。代码如下所示。

```
>>> import pandas as pd
>>> df = pd.read_excel("F:\\data\shares.xlsx","Sheet1")
>>> df.columns
Index(['share','quantity','price','description','amount invested','profit%'],
dtype='object')
```

假设只想检索 description 列。
从图 7.42 中可以看到，要仅检索 description 列的前 5 个条目，可以使用以下代码：

```
>>> df.description.head()
```

要检索 description 列的所有条目，只需使用以下代码：

```
>>> df.description
```

输出结果如图 7.42 所示。

```
>>> df.description.head()
0                                              POWER
1                                                NaN
2       4th largest manufacturer of thin polyester Film
3                                                NaN
4                                                NaN
Name: description, dtype: object
>>> df.description
0                                              POWER
1                                                NaN
2       4th largest manufacturer of thin polyester Film
3                                                NaN
4                                                NaN
5                                                NaN
6                                                NaN
7                                                NaN
8                                                NaN
9                                                NaN
10                                               NaN
11                                               NaN
12                                               NaN
13                                               NaN
14                                               NaN
15                                               NaN
16                                               NaN
17                                               NaN
18                                               NaN
19                                               NaN
20                                               NaN
21                                               NaN
22                                               NaN
23                                               NaN
24                                               NaN
25                                               NaN
26                                               NaN
27                                               NaN
```

图 7.42

还有一种访问列内容的方法。假设想查看 amount invested 列。由于这两个单词之间有空格，得到如下结果：

```
>>> df['amount invested']
```

使用 head() 函数检索 amount invested 列中前 5 条的代码如下所示。

```
>>> df['amount invested'].head()
```

输出结果如图 7.43 所示。

```
>>> df['amount invested']
0      3740.40
1      1566.08
2      4840.00
3      1340.00
4      2399.25
5      1877.50
6      8269.80
7      4383.00
8      3823.00
9      3867.50
10      773.50
11      504.14
12     4158.70
13     1097.00
14     2571.90
15     1104.50
16         NaN
17      936.90
18      972.00
19      483.60
20      774.25
21     2342.10
22         NaN
23         NaN
24         NaN
25         NaN
26         NaN
27    51825.12
Name: amount invested, dtype: float64
>>> df['amount invested'].head()
0    3740.40
1    1566.08
2    4840.00
3    1340.00
4    2399.25
Name: amount invested, dtype: float64
>>>
```

图 7.43

类似地，使用以下方法检索 amount invested 列中的最后 5 条，输出结果如图 7.44 所示。

```
df['amount invested'].tail()
```

```
>>> df['amount invested'].tail()
23         NaN
24         NaN
25         NaN
26         NaN
27    51825.12
Name: amount invested, dtype: float64
>>>
```

图 7.44

3. 检索多列数据

使用以下方法检索多列数据：

```
DataFrame_name.[[col_1,col_2,…]]
```

输出结果如图 7.45 所示。

```
>>> df[['price','amount invested']]
      price   amount invested
0    311.70           3740.40
1     97.88           1566.08
2    484.00           4840.00
3     33.50           1340.00
4    479.85           2399.25
5    187.75           1877.50
6   2756.60           8269.80
7   2191.50           4383.00
8   1911.50           3823.00
9    773.50           3867.50
10    59.50            773.50
11    38.78            504.14
12   415.87           4158.70
13   109.70           1097.00
14   857.30           2571.90
15   110.45           1104.50
16      NaN               NaN
17   104.10            936.90
18   108.00            972.00
19    37.20            483.60
20   154.85            774.25
21   780.70           2342.10
22   786.25               NaN
23  1727.55               NaN
24   722.10               NaN
25  1636.35               NaN
26   713.40               NaN
27      NaN          51825.12
>>>
```

图 7.45

head()函数和tail()函数的用法相同,如图7.46所示。

```
>>> df[['price','amount invested']].tail(2)
    price   amount invested
26  713.4               NaN
27    NaN          51825.12
```

图 7.46

使用 max() 函数和 min() 函数计算最大值和最小值。代码如下所示。

```
>>> df.price.max()
2756.6
>>> df['amount invested'].min()
483.6
>>> df['profit%'].max()
0.21
```

使用 describe() 函数显示信息,如图 7.47 所示。

```
>>> df.describe()
       quantity       price  amount invested    profit%
count  26.000000   26.000000        22.000000  13.000000
mean    8.230769  676.533846      4711.374545   0.123846
std     7.900925  755.414971     10692.573093   0.057813
min     1.000000   33.500000       483.600000   0.020000
25%     3.000000  108.425000      1003.250000   0.080000
50%     7.500000  447.860000      2109.800000   0.140000
75%    10.000000  784.862500      3856.375000   0.150000
max    40.000000 2756.600000     51825.120000   0.210000
>>>
```

图 7.47

describe()函数可以提供以下信息。

(1) 值总计数；
(2) 平均数；
(3) 标准差；
(4) 最小值；
(5) 最大值；
(6) 总值的25%、50%和75%。

4. 根据条件检索数据

示例 7.10

检索利润百分比最大时的股票、描述、价格值。

代码：

```
>>> import pandas as pd
>>> import xlrd
>>> df = pd.read_excel("F:\\data\shares.xlsx","Sheet1")
>>> #检索列名称
>>> df.columns
Index(['share','quantity','price','description','amount invested','profit%'],
dtype='object')
>>> #检索利润百分比最大时的股票、描述、价格值
>>> df[['share','description','price']][df['profit%']==df['profit%'].max()]
        share description   price
20  tatamotors     NaN    154.85
```

示例 7.11

检索价格值最小时的股票、投资金额、价格值。

代码：

```
>>> #检索价格值最小时的股票、投资金额、价格值
>>> df[['share','amount invested','profit%']][df['profit']==df['profit'].min()]
        share  amount invested  profit%
19  tatamotors          774.25     0.21
>>>
```

示例 7.12

检索投资金额超过 500 卢比的股票。

代码：

```
>>> [['share']][df['amount invested']>=500]
        share
0   Coal india
1   Powergrid
```

```
2      Polyplex Corporation
3             noida toll
4             Kotak bank
5                   Sail
6                    SBI
7                    SBI
8                    SBI
9               Reliance
10          JP Associate
11                suzlon
12            tata steel
13                  SAIL
14                 ICICI
15                  IDFC
16                  SAIL
17                  IDFC
18            tatamotors
19                 ICICI
20                 Total
```

5. 索引范围

使用以下方法检索索引信息：

```
>>> import pandas as pd
>>> df = pd.read_excel("F:\\data\shares.xlsx","Sheet1")
>>> df.index
RangeIndex(start=0, stop=28, step=1)
```

从本章开头知道索引列可以自动生成。任意具有唯一值的列都可以设置为索引。添加名为 share_id 的列，如图 7.48 所示。

	A	B	C	D	E	F	G
1	share_id	share	quantity	price	description	amount invested	profit%
2	sh001	Coal india	12	311.7	POWER	3740.4	15%
3	sh002	Powergrid	16	97.88		1566.08	7%
4	sh003	Polyplex Corporation	10	484	4th largest manufacturer of thin polyester Film	4840	
5	sh004	noida toll	40	33.5		1340	
6	sh005	Kotak bank	5	479.85		2399.25	2.00%
7	sh006	Sail	10	187.75		1877.5	4%
8	sh007	SBI	3	2756.6		8269.8	
9	sh008	SBI	2	2191.5		4383	
10	sh009	SBI	2	1911.5		3,823	
11	sh010	Reliance	5	773.5		3867.5	11%
12	sh011	JP Associate	13	59.5		773.5	19%
13	sh012	suzlon	13	38.78		504.14	
14	sh013	tata steel	10	415.87		4158.7	14%
15	sh014	SAIL	10	109.7		1097	
16	sh015	ICICI	3	857.3		2571.9	8%
17	sh016	IDFC	10	110.45		1104.5	12%
18							15%

图 7.48

share_id 列有唯一值，可以将其设置为索引。

代码：

```
>>> import pandas as pd
>>> df = pd.read_excel("F:\\data\shares.xlsx","Sheet1")
>>> df.index
RangeIndex(start =0, stop =28, step =1)
>>> df1 = df.set_index('share_id')
>>> df1
```

输出结果如图 7.49 所示。

```
>>> df1
              share         quantity  ...  amount invested  profit%
share_id                               ...
sh001         Coal india        12.0  ...          3740.40     0.15
sh002         Powergrid         16.0  ...          1566.08     0.07
sh003  Polyplex Corporation    10.0  ...          4840.00      NaN
sh004         noida toll       40.0  ...          1340.00      NaN
sh005         Kotak bank        5.0  ...          2399.25     0.02
sh006               Sail       10.0  ...          1877.50     0.04
sh007                SBI        3.0  ...          8269.80      NaN
sh008                SBI        2.0  ...          4383.00      NaN
sh009                SBI        2.0  ...          3823.00      NaN
sh010            Reliance       5.0  ...          3867.50     0.11
sh011        JP Associate      13.0  ...           773.50     0.19
sh012              suzlon      13.0  ...           504.14      NaN
sh013          tata steel      10.0  ...          4158.70     0.14
sh014                SAIL      10.0  ...          1097.00      NaN
sh015               ICICI       3.0  ...          2571.90     0.08
sh016                IDFC      10.0  ...          1104.50     0.12
sh017                SAIL       9.0  ...           936.90      NaN
sh018                IDFC       9.0  ...           972.00     0.15
sh019              SUZLON      13.0  ...           483.60      NaN
sh020          tatamotors       5.0  ...           774.25     0.21
sh021               ICICI       3.0  ...          2342.10     0.18
sh022               icici       6.0  ...              NaN      NaN
sh023                 sbi       2.0  ...              NaN      NaN
sh024               icici       1.0  ...              NaN      NaN
sh025                 sbi       1.0  ...              NaN      NaN
sh026               icici       1.0  ...              NaN      NaN
sh027               Total       NaN  ...         51825.12      NaN

[27 rows x 6 columns]
```

图 7.49

这时，dataframe df 的内容保持不变。

使用 loc 属性找到共享内容。

假设在索引 2 处查找共享信息。代码如下所示。

```
>>> df.loc[2]
share_id                                                      sh003
share                                          Polyplex Corporation
quantity                                                         10
price                                                           484
description       4th largest manufacturer of thin polyester Film
amount invested                                                4840
profit%                                                         NaN
Name: 2, dtype: object
```

使用 df1 检索相同信息。代码如下所示。

```
>>> df1.loc['sh003']
share                        Polyplex Corporation
quantity                                       10
price                                         484
description       4th largest manufacturer of thin polyester Film
amount invested                              4840
profit%                                       NaN
Name: sh003, dtype: object
```

使用以下指令将 share_id 列设为 df 的索引：

```
>>> import pandas as pd
>>> df = pd.read_excel("F:\\data\shares.xlsx","Sheet1")
>>> df.set_index('share_id',inplace = True)
>>> df
```

输出结果如图 7.50 所示。

```
          share              quantity  ...  amount invested  profit%
share_id                                ...
sh001           Coal india       12.0  ...          3740.40     0.15
sh002            Powergrid       16.0  ...          1566.08     0.07
sh003  Polyplex Corporation      10.0  ...          4840.00      NaN
sh004           noida toll       40.0  ...          1340.00      NaN
sh005           Kotak bank        5.0  ...          2399.25     0.02
sh006                 Sail       10.0  ...          1877.50     0.04
sh007                  SBI        3.0  ...          8269.80      NaN
sh008                  SBI        2.0  ...          4383.00      NaN
sh009                  SBI        2.0  ...          3823.00      NaN
sh010             Reliance        5.0  ...          3867.50     0.11
sh011          JP Associate      13.0  ...           773.50     0.19
sh012               suzlon       13.0  ...           504.14      NaN
sh013           tata steel       10.0  ...          4158.70     0.14
sh014                 SAIL       10.0  ...          1097.00      NaN
sh015                ICICI        3.0  ...          2571.90     0.08
sh016                 IDFC       10.0  ...          1104.50     0.12
sh017                 SAIL        9.0  ...           936.90      NaN
sh018                 IDFC        9.0  ...           972.00     0.15
sh019               SUZLON       13.0  ...           483.60      NaN
sh020           tatamotors        5.0  ...           774.25     0.21
sh021                ICICI        3.0  ...          2342.10     0.18
sh022                icici        6.0  ...              NaN      NaN
sh023                  sbi        2.0  ...              NaN      NaN
sh024                icici        1.0  ...              NaN      NaN
sh025                  sbi        1.0  ...              NaN      NaN
sh026                icici        1.0  ...              NaN      NaN
sh027                Total        NaN  ...         51825.12      NaN

[27 rows x 6 columns]
```

图 7.50

6. 重置索引

重置 df 索引。

代码：

```
>>> df.reset_index(inplace = True)
>>> df
```

输出结果如图 7.51 所示。

```
>>> df.reset_index(inplace = True)
>>> df
   share_id                share  ...  amount invested  profit%
0     sh001           Coal india  ...          3740.40     0.15
1     sh002            Powergrid  ...          1566.08     0.07
2     sh003  Polyplex Corporation ...          4840.00      NaN
3     sh004           noida toll  ...          1340.00      NaN
4     sh005           Kotak bank  ...          2399.25     0.02
5     sh006                 Sail  ...          1877.50     0.04
6     sh007                  SBI  ...          8269.80      NaN
7     sh008                  SBI  ...          4383.00      NaN
8     sh009                  SBI  ...          3823.00      NaN
9     sh010             Reliance  ...          3867.50     0.11
10    sh011         JP Associate  ...           773.50     0.19
11    sh012               suzlon  ...           504.14      NaN
12    sh013           tata steel  ...          4158.70     0.14
13    sh014                 SAIL  ...          1097.00      NaN
14    sh015                ICICI  ...          2571.90     0.08
15    sh016                 IDFC  ...          1104.50     0.12
16    sh017                 SAIL  ...           936.90      NaN
17    sh018                 IDFC  ...           972.00     0.15
18    sh019               SUZLON  ...           483.60      NaN
19    sh020           tatamotors  ...           774.25     0.21
20    sh021                ICICI  ...          2342.10     0.18
21    sh022                icici  ...              NaN      NaN
22    sh023                  sbi  ...              NaN      NaN
23    sh024                icici  ...              NaN      NaN
24    sh025                  sbi  ...              NaN      NaN
25    sh026                icici  ...              NaN      NaN
26    sh027                Total  ...         51825.12      NaN

[27 rows x 7 columns]
>>>
```

图 7.51

7. 排序数据

使用 sort_values( ) 函数对数据进行排序。

代码：

```
>>> df.sort_values('profit%')
```

输出结果如图7.52所示。

```
>>> df.sort_values('profit%')
    share_id              share  ...  amount invested  profit%
4      sh005         Kotak bank  ...          2399.25     0.02
5      sh006               Sail  ...          1877.50     0.04
1      sh002          Powergrid  ...          1566.08     0.07
14     sh015              ICICI  ...          2571.90     0.08
9      sh010           Reliance  ...          3867.50     0.11
15     sh016               IDFC  ...          1104.50     0.12
12     sh013         tata steel  ...          4158.70     0.14
0      sh001        Coal india  ...          3740.40     0.15
17     sh018               IDFC  ...           972.00     0.15
20     sh021              ICICI  ...          2342.10     0.18
10     sh011       JP Associate ...           773.50     0.19
19     sh020         tatamotors  ...           774.25     0.21
2      sh003  Polyplex Corporation ...        4840.00      NaN
3      sh004         noida toll  ...          1340.00      NaN
6      sh007                SBI  ...          8269.80      NaN
7      sh008                SBI  ...          4383.00      NaN
8      sh009                SBI  ...          3823.00      NaN
11     sh012             suzlon  ...           504.14      NaN
13     sh014               SAIL  ...          1097.00      NaN
16     sh017               SAIL  ...           936.90      NaN
18     sh019             SUZLON  ...           483.60      NaN
21     sh022              icici  ...              NaN      NaN
22     sh023                sbi  ...              NaN      NaN
23     sh024              icici  ...              NaN      NaN
24     sh025                sbi  ...              NaN      NaN
```

图 7.52

输入以下代码获得最高利润份额:

```
df.sort_values('profit%', ascending = False)
```

输出结果如图7.53所示。

```
>>> df.sort_values('profit%', ascending = False)
    share_id              share  ...  amount invested  profit%
19     sh020         tatamotors  ...           774.25     0.21
10     sh011       JP Associate ...           773.50     0.19
20     sh021              ICICI  ...          2342.10     0.18
0      sh001         Coal india  ...          3740.40     0.15
17     sh018               IDFC  ...           972.00     0.15
12     sh013         tata steel  ...          4158.70     0.14
15     sh016               IDFC  ...          1104.50     0.12
9      sh010           Reliance  ...          3867.50     0.11
14     sh015              ICICI  ...          2571.90     0.08
1      sh002          Powergrid  ...          1566.08     0.07
5      sh006               Sail  ...          1877.50     0.04
4      sh005         Kotak bank  ...          2399.25     0.02
2      sh003  Polyplex Corporation ...        4840.00      NaN
3      sh004         noida toll  ...          1340.00      NaN
6      sh007                SBI  ...          8269.80      NaN
7      sh008                SBI  ...          4383.00      NaN
8      sh009                SBI  ...          3823.00      NaN
11     sh012             suzlon  ...           504.14      NaN
13     sh014               SAIL  ...          1097.00      NaN
```

图 7.53

小结　　无论何时,都必须不断提高自己,这样才能进步。只有意识到自身优势和劣势,才能提高自己。企业也如此。依据数据信息能确定企业的优势和劣势。在实际场景中要处理的数据量巨大,将数据转换为图有助于理解实际情况。本章介绍了Python数据可视化和分析的各种工具。

第 8 章

创建 GUI 表和添加控件

> **引言**　本章介绍如何使用 tkinter 模块。Python 使用 tkinter 模块创建 GUI，该模块是处理图形的工具。

知识结构

- 开始
- 控件
- 按钮和消息框
- Canvas
- Frame
- 标签
- 小项目——秒表
- 列表框
- 菜单按钮和菜单
- 单选按钮
- 滚动条和滑块
- 文本框
- Spinbox

目标

完成本章的学习后，读者应掌握以下技能。

（1）使用 tkinter 模块和各种控件。

（2）创建具有基本功能的 GUI 应用程序。

tkinter 是使用工具命令语言（TCL）的 TK 模块类，该模块以适用于 Web 和桌面应用程序以及网络、管理等著称。TK 表示工具包，TCL 使用 TK 创建图形。TK 提供标准 GUI，能被 Python 等动态编程语言使用。Python 程序开发人员能使用 tkinter 模块访问 TK。Python 提供多种开发 GUI 的方法。创建 GUI 最简单快捷的方法是使用 Python 和 tkinter。因此，tkinter 是最常用的 GUI 模块。

8.1 开始

在使用 GUI 之前，需要导入 tkinter 模块。

这是使用 GUI 所有代码的第一步，可以使用以下语句：

```
from tkinter import *
```

或

```
import tkinter as t
```

示例 8.1

编写代码创建主窗口对象。

代码：

```
>>> import tkinter
>>> mw = Tk()
```

或

```
>>> import tkinter as t
>>> # 创建名为 mw 的主窗口(实际上可以是任何东西)
>>> mw = t.Tk()
```

这将打开主窗口。

输出结果如图 8.1 所示。

图 8.1

现在介绍非常重要的 mainloop( )方法,运行应用程序时必须对其进行调用。mainloop( )方法是无限循环,用于运行应用程序。只要窗口未关闭,就等待事件并响应该事件。

启动简单窗口代码如下所示。

```
>>> import tkinter
>>> mw = tkinter.Tk()
>>> mw.mainloop()
```

或

```
>>> import tkinter as t
>>> mw = t.Tk()
>>> mw.mainloop()
```

示例 8.2

编写代码设置主窗口标题。

使用 title( )方法设置主窗口标题。

代码:

```
#导入 tkinter
from tkinter import *

#创建主窗口
mw = Tk()

#设置主窗口标题
mw.title("my First window")

mw.mainloop()
```

输出结果如图 8.2 所示。

图 8.2

窗口非常小。使用 geometry( ) 函数设置窗口大小。

示例 8.3
编写代码设置窗口大小。
代码：

```
#导入 tkinter
from tkinter import *

#创建主窗口
mw = Tk()

#设置主窗口标题
mw.title("my First window")

#设置窗口大小
mw.geometry("600x400")

mw.mainloop()
```

输出结果如图 8.3 所示。

图 8.3

示例 8.4

编写程序改变主窗口图像。

现在更改主窗口的叶子图像,使用下面存储在目录中的水滴图像(见图 8.4),图像文件存储在"F:/my_images"文件夹中。

用新图像替换叶子图像,图像文件只能是".ico"格式。使用 wm.iconbitmap( )语句实现。

图 8.4

代码:

```
#导入 tkinter
from tkinter import *

#创建主窗口
mw = Tk()

#设置主窗口标题
mw.title("my First window")

#设置窗口大小
mw.geometry("600x400")

#设置窗口的图像图标(在本例中,图像位于F:/my_images/drop.ico)

#提供图像位置
mw.wm_iconbitmap("F:/my_images/drop.ico")

mw.mainloop()
```

输出结果如图 8.5 所示。

图 8.5

8.2 控件

本节使用标准 GUI 元素（如按钮、标签等）等控件。使用控件需执行以下步骤。
步骤 1：导入 tkinter。
步骤 2：创建主窗口。
步骤 3：设置主窗口标题。
步骤 4：编写控件处理代码。
步骤 5：调用 mainloop() 方法响应用户触发事件。
代码的基本结构如下所示。

```
#导入tkinter
from tkinter import *

# 创建主窗口
mw = Tk()

# 设置主窗口标题(可选项)
mw.title("my First window")

# 编写控件处理代码
-------------- Your code comes here

# 调用mainloop()方法
mw.mainloop()
```

图 8.6 所示为一些重要控件。

图 8.6

1. 布局管理

布局管理（也称为几何管理）在 GUI 应用程序设计中非常重要。tkinter 模块提供 3

种方法进行布局管理，它们不能同时在同一个父窗口中使用。这 3 种方法为 pack( )、grid( ) 和 place( )。

使用上面 3 种布局管理方法将控件放在父窗口中，安排控件在屏幕中的位置。添加控件时确定控件大小，但控件在父窗口中的位置取决于布局管理设置。

1）pack( ) 方法

pack( ) 方法最容易使用，本书中大多数示例都使用 pack( ) 方法。pack( ) 方法无须专门定义控件必需的显示位置，但必须相对于其他控件声明该位置。虽然 pack( ) 方法是最容易使用的布局管理方法，但 grid( ) 和 place( ) 方法更具灵活性。

语法格式如下所示。

```
widget_name.pack(options)
```

假设有 3 个按钮：my_first_button、my_second_button 和 my_third_button。

（1）使用没有参数的 pack( ) 方法布局按钮。

代码：

```
my_first_button.pack()
my_second_button.pack()
my_third_button.pack()
```

输出结果如图 8.7 所示。

图 8.7

（2）使用 pack(expand) 布局按钮。

其中，参数 expand 可以取两个值：True 或 False。

代码：

```
my_first_button.pack(expand = True)
my_second_button.pack(expand = True)
my_third_button.pack(expand = True)
```

输出结果如图 8.8 所示。

图 8.8

(3) 使用 pack(fill) 布局按钮。

其中, 参数 fill 的取值为 x、y、both、none。

输出结果如图 8.9 所示。

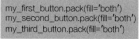

图 8.9

(4) 使用 pack(side) 布局按钮。

其中, 参数 side 可以控制控件出现在父窗口哪一侧。其可取的 4 个值分别是 top、bottom、left、right。

代码:

```
my_first_button.pack(side = 'top')
my_second_button.pack(side = 'bottom')
my_third_button.pack(side = 'left')
```

输出结果如图 8.10 所示。

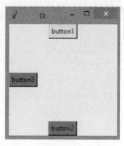

图 8.10

2) grid()方法

grid()是另一种以类似表格的结构组织控件的方法,也是布局几何图形的首选方法之一。grid()方法通常与对话框一起使用,根据网格位置坐标定位控件。grid()方法如表 8.1 所示。

表 8.1

grid()方法	说明
column, row	确定控件的坐标
ipadx, ipady	控件内框水平和垂直填充量
padx, pady	控件外框水平和垂直填充量
Columnspan	控件占用的列数,默认值为 1
Rowspan	控件占用的行数,默认值为 1
sticky	小于单元格时定义控件粘贴位置 ● W——靠左粘贴 ● E——靠右粘贴 ● N——靠上粘贴 ● S——靠下粘贴

代码:

```
import tkinter as tk

mw = tk.Tk()

labelusern = tk.Label(mw, text = "User Name")
labelusern.grid(column = 0, row = 0, ipadx = 5, pady = 5, sticky = tk.W + tk.N)

labelpwd  = tk.Label(mw, text = "Password")
labelpwd.grid(column = 0, row = 1, ipadx = 5, pady = 5, sticky = tk.W + tk.S)

entryusern = tk.Entry(mw, width = 20)
entrypwd = tk.Entry(mw, width = 20)

entryusern.grid(column = 1, row = 0, padx = 10, pady = 5, sticky = tk.N)
entrypwd.grid(column = 1, row = 1, padx = 10, pady = 5, sticky = tk.S)

loginButton = tk.Button(mw, text = 'Login')
loginButton.grid(column = 0, row = 2, pady = 10, sticky = tk.W)

mw.mainloop()
```

输出结果如图 8.11 所示。

图 8.11

3) place() 方法

使用 place() 方法可以将窗口控件以绝对或相对方式放置在窗口中。以下代码是绝对位置示例。

```python
import tkinter as tk

mw = tk.Tk()

#定义控件
labelusern = tk.Label(mw, text = "User Name")
labelpwd = tk.Label(mw,text = "Password")
entryusern = tk.Entry(mw, width =20)
entrypwd = tk.Entry(mw, width =20)
loginButton = tk.Button(mw, text = 'Login')

#设置控件位置
labelusern.place(x = 10, y = 5)
labelpwd.place(x = 10, y = 40)
entryusern.place(x = 75, y = 5)
entrypwd.place(x = 75, y = 40)
loginButton.place(x = 10, y = 80)
mw.mainloop()
```

输出结果如图 8.12 所示。

图 8.12

2. 相对位置

在下面的代码中，注意 relx 和 rely 是控件位置与窗口大小的相对百分比。例如，

relx=0.01，rely=0.15，表示控件放置在窗口宽度1%和窗口高度15%处。

代码：

```python
import tkinter as tk

mw = tk.Tk()

#定义控件
labelusern = tk.Label(mw, text = "User Name")
labelpwd = tk.Label(mw,text = "Password")
entryusern = tk.Entry(mw, width=20)
entrypwd = tk.Entry(mw, width=20)
loginButton = tk.Button(mw, text = 'Login')

#设置控件位置
labelusern.place(relx = 0.01, rely = 0.15)
labelpwd.place(relx = 0.01, rely = 0.35)
entryusern.place(relx = 0.35, rely = 0.15)
entrypwd.place(relx = 0.35, rely = 0.35)
loginButton.place(relx = 0.01, rely = 0.55)
mw.mainloop()
```

输出结果如图8.13所示。

图8.13

8.3 按钮和消息框

按钮是常用的用户交互控件，单击会触发事件，该事件通常定义在单击按钮时调用的函数中。可以在按钮上放置文本或图像描述其用途。

语法格式如下所示。

```
button_name = Button(mw, option = value)
```

（1）mw：父窗口，在这个示例中是主窗口。

（2）option = value：这是键-值对，用逗号分隔，如背景颜色、字体、图像、边框等。

详细情况参见表 8.2。

表 8.2

选项	描述	代码	输出结果
activeback-ground	鼠标经过时按钮的颜色	my_first_button = Button(mw, activebackground="yellow", text="Click Here")	activebackground = "yellow"
activefore-ground	鼠标经过时按钮的文本颜色	my_first_button = Button(mw, activeforeground="red", text="Click Here", command=message_display)	activeforeground = "red"
bd	边框宽度，默认值为 2	my_first_button = Button(mw, bd="10", text="Click Here")	bd = "10"
bg	背景色	my_first_button = Button(mw, bg="skyblue", text="Click Here")	bg = "skyblue"
command	单击按钮时要执行的操作	my_first_button = Button(mw, text="Click Here", command=message_display)	

续表

选项	描述	代码	输出结果
fg	正常的前景/文本颜色	`my_first_button = Button(mw, fg = "purple", text = "Click Here")`	**fg = "purple"**
font	按钮上使用的文本字体	`my_first_button = Button(mw, font = "Helvetica 10 bold italic", text = "Click Here")`	**font = "Helvetica 10 bold italic"**
height	按钮高度	`my_first_button = Button( mw, height = "6", text = "Click Here")`	**height = "6"**
pady	文本上方和下方的附加填充	`my_first_button = Button( mw, text = "Click Here", pady = "15")`	**pady = "15"**
relief	定义边框类型：平、凹、凸、脊状、实心和凹陷的	`my_first_button = Button( mw, text = "Click Here", relief = "sunken")`	**relief = "sunken"**

续表

选项	描述	代码	输出结果
state	按钮状态：激活、禁用、正常	`my_first_button = Button(mw, text = "Click Here", state = "disabled")`	state = "disabled"
underline	如果是-1，则字符无下划线。如果是非负数，则相应字符有下划线	`my_first_button = Button(mw, text = "Click Here", underline = "4")`	underline = "4"

下面介绍如何编写显示按钮的代码。具体步骤如下。

步骤1：导入 tkinter 获得 TK 的所有函数。代码如下所示。

```
from tkinter import *
```

步骤2：创建窗口实例。代码如下所示。

```
mw = Tk()
```

步骤3：设置窗口大小。代码如下所示。

```
mw.geometry("150x150")
```

步骤4：创建按钮实例。代码如下所示。

```
my_first_button = Button(mw, text = "Click Here")
```

其中，参数 mw 表示按钮关联的窗口实例，参数 text 表示按钮的显示文本，但不会在窗口上显示按钮。

步骤5：使用 pack() 方法将按钮显示到窗口上。代码如下所示。

```
my_first_button.pack()
```

步骤6：调用 mainloop() 方法。代码如下所示。

```
mw.mainloop()
```

代码：

```
#导入tkinter 获得 Tk 的所有函数
from tkinter import *

#创建窗口实例
mw = Tk()

#设置窗口大小
mw.geometry("150x150")

#创建按钮实例
my_first_button = Button(mw, text = "Click Here")

#使用pack()方法将按钮显示到窗口上
my_first_button.pack()

#调用 mainloop()方法
mw.mainloop()
```

输出结果如图 8.14 所示。

图 8.14

在前面示例中，只是简单显示按钮，尚未定义该按钮的用途或关联代码。

示例 8.5

编写代码，单击按钮时，显示带有消息"welcome to the world of Widgets"的消息框。步骤如下。

步骤 1：导入语句。

单击按钮时显示消息框。因此，必须导入 tkinter.messagebox，注意是模块而不是类。代码如下所示。

```
from tkinter import *
import tkinter.messagebox
```

步骤 2：定义单击按钮时的调用函数。

showinfo()函数有 2 个参数：第 1 个参数是字符串，表示窗口标题；第 2 个参数也是字符串，表示显示信息。代码如下所示。

```
def message_display():
    tkinter.messagebox.showinfo("Welcome Note","welcome to the world of Widgets")
```

步骤 3：创建窗口实例。代码如下所示。

```
mw = Tk()
```

步骤 4：设置窗口大小。代码如下所示。

```
mw.geometry("150x150")
```

步骤 5：创建按钮实例。

Button()函数有 3 个参数：第 1 个参数 mw 是按钮关联的窗口实例；第 2 个参数 text 是按钮显示文本；第 3 个参数 command 是单击按钮时响应的操作命令。此时，要调用步骤 2 中的函数。注意参数 command 可以是函数、绑定方法或任意可调用的 Python 对象。如果不使用该选项，用户按下按钮时什么也不会发生。代码如下所示。

```
my_first_button = Button(mw, text = "Click Here", command = message_display)
```

步骤 6：显示按钮，将按钮布局到窗口中。代码如下所示。

```
my_first_button.pack()
```

步骤 7：调用 mainloop()方法。代码如下所示。

```
mw.mainloop()
```

代码：

```
#导入语句
from tkinter import *
import tkinter.messagebox

#定义单击按钮时的调用函数
def message_display():
    tkinter.messagebox.showinfo("Welcome Note","welcome to the world of Widgets")

#创建窗口实例
mw = Tk()

#设置窗口大小
mw.geometry("150x150")

#创建按钮实例
my_first_button = Button(mw, text = "Click Here", command = message_display)

#将按钮布局到窗口中
my_first_button.pack()

#调用 mainloop()方法
mw.mainloop()
```

输出结果如图 8.15 所示。

图 8.15

单击该按钮出现图 8.16 所示消息。

图 8.16

示例 8.6

编写创建两个按钮的程序：左右各一个。

代码：

```
#导入语句
from tkinter import *
import tkinter.messagebox

# 创建右按钮函数
def message_display_right():
    tkinter.messagebox.showinfo("Next Topic","Welcome to Canvas")

# 创建左按钮函数
def message_display_left():
    tkinter.messagebox.showinfo("Previous Topic","Welcome to Widgets")

#创建窗口实例
mw = Tk()
mw.title("Select Topic")

#创建按钮实例
my_first_button = Button(mw,text = "Next",fg = "Green", command = message_display_right)
my_second_button = Button(mw,text = "Previous",fg = "Red", command = message_display_left)
```

```
#调整按钮位置
my_first_button.pack(side = tkinter.RIGHT)
my_second_button.pack(side = tkinter.LEFT)

#调用 mainloop()方法
mw.mainloop()
```

输出结果如图 8.17 所示。

图 8.17

单击"Next"按钮,结果如图 8.18 所示。

图 8.18

单击"Previous"按钮,结果如图 8.19 所示。

图 8.19

示例 8.7

编写代码,向按钮中添加图像(从 f:\\my_images\\thumb2.gif 获得)。
代码:

```
#导入语句
from tkinter import *
import tkinter.messagebox
from tkinter import ttk

# 创建右按钮函数
def message_display():
    tkinter.messagebox.showinfo ( " Experimenting with GIFs "," Welcome to
    GIFs")
```

```
#创建窗口实例
mw = Tk()
mw.title("Thumbs UP")

#创建按钮实例
my_first_button = ttk.Button(mw,text = "click", command = message_display)
my_first_button.pack()

#添加按钮图像
img = PhotoImage(file = "f:\\my_images\\thumb2.gif")
my_first_button.config(image = img,compound =RIGHT)

#调用mainloop()方法
mw.mainloop()
```

输出结果如图 8.20 所示。

图 8.20

单击"click"按钮,结果如图 8.21 所示。

图 8.21

8.4 Canvas

本节介绍 Canvas,其用于绘制图像的矩形区域。语法格式如下所示。

```
c = Canvas(master, option = value…)
```

其中,c 是 Canvas 类对象;master 是主窗口;option 是 Canvas 对象的常用选项。所有 option = value 对都用逗号隔开。

在开始使用 Canvas 之前,先看基本步骤,将 Canvas 关联到主窗口。

步骤 1:导入 tkinter。代码如下所示。

```
from tkinter import *
```

步骤2：创建窗口实例。代码如下所示。

```
mw = Tk()
```

步骤3：创建 Canvas 类实例。

这里，使用 200 像素×200 像素大小。可以根据实际情况赋予任何值。代码如下所示。

```
mc = Canvas(mw, width = 200, height = 200)
```

步骤4：在窗口对象 mw 上添加 Canvas 对象 mc。代码如下所示。

```
mc.pack()
```

步骤5：调用 mainloop() 方法。代码如下所示。

```
mw.mainloop()
```

代码：

```
#导入 tkinter
from tkinter import *

#创建窗口实例
mw = Tk()

#创建 Canvas 类实例
mc = Canvas(mw, width = 200, height = 200)

#在窗口对象 mw 上添加 Canvas 对象 mc
mc.pack()

#调用 mainloop() 方法
mw.mainloop()
```

输出结果如图 8.22 所示。

图 8.22

下面分析输出窗口。

图 8.23 中间的标记区域是可绘制 Canvas。x 轴正向向右，y 轴正向向下。Canvas 尺寸为 200 像素×200 像素。其左上角为（x = 0，y = 0），右上角为（x = 200，y = 0）。同理，Canvas 的左下角为（x = 0，y = 200），右下角为（x = 200，y = 200）。

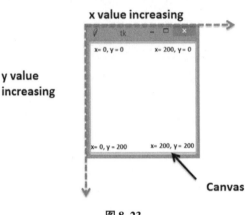

图 8.23

现在绘制连接 Canvas 左上角和右下角的线。

语法格式如下所示。

```
line = Canvas.create_line(x0, y0, x1, y1,…, xn, yn, other options)
```

选项 fill 用于设置在 Canvas 上绘制蓝线。只需在前面示例中添加一行代码。

代码：

```
#导入 tkinter
from tkinter import *

#创建窗口实例
mw = Tk()

#创建 Canvas 类实例
mc = Canvas(mw, width = 200, height = 200)

#绘制线
line = mc.create_line(0,0,200,200,fill = "Blue")

#在窗口对象 mw 上添加 Canvas 对象 mc
mc.pack()

#调用 mainloop()方法
mw.mainloop()
```

输出结果如图 8.24 所示。

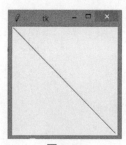

图 8.24

示例 8.8

连接图 8.25 中的点 2、点 3 和点 4。

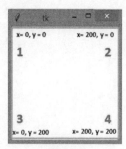

图 8.25

观察下面的点。

（1）点 2(200,0)连接到点 3(0,200)；
（2）点 3(0,200)连接到点 4(200,200)；
（3）点 4(200,200)连接到点 2(200,0)。

因此，有：

（1）（x0，y0）=（200,0）；
（2）（x1，y1）=（0,200）；
（3）（x2，y2）=（200,200）；
（4）（x3，y3）=（200,0）。

结果如图 8.26 所示。

line = mc.create_line(200,0,0,200,200,200,200,0,fill = "orange")

图 8.26

代码：

```
#导入tkinter
from tkinter import *

#创建窗口实例
mw = Tk()
```

```
#创建 Canvas 类实例
mc = Canvas(mw, width = 200, height = 200)

#创建连接点 2、3 和 4 的线
line = mc.create_line(200,0,0,200,200,200,0,fill = "orange")

#在窗口对象 mw 上添加 Canvas 对象 mc
mc.pack()

#调用 mainloop()方法
mw.mainloop()
```

输出结果如图 8.27 所示。

图 8.27

编程实现在 Canvas 中绘制十字，线宽为 10。

代码：

```
#导入 tkinter
from tkinter import *

#创建窗口实例
mw = Tk()

#创建 Canvas 类实例
mc = Canvas(mw, width = 200, height = 200)

#绘制线
line = mc.create_line(200,0,0,200,fill = "Purple",width = 10)
line2 = mc.create_line(0,0,200,200,fill = "Blue",width = 10)

#在窗口对象 mw 上添加 Canvas 对象 mc
mc.pack()

#调用 mainloop()方法
mw.mainloop()
```

输出结果如图 8.28 所示。

图 8.28

下面快速学习在 Canvas 中绘制其他形状的方法。

1. 矩形

以下示例显示如何绘制矩形。

示例 8.9

编写代码绘制矩形，该矩形在图形中间框中突出显示。

代码：

```
#导入 tkinter
from tkinter import *

#创建窗口实例
mw = Tk()

#创建 Canvas 类实例
mc = Canvas(mw, width = 200, height = 200)

#绘制矩形
rect = mc.create_rectangle(5,10,100,50,fill = "Purple", width = 10)

#在窗口对象 mw 上添加 Canvas 对象 mc
mc.pack()

#调用 mainloop()方法
mw.mainloop()
```

输出结果如图 8.29 所示。

图 8.29

分析:

如图 8.30 所示,矩形左上角的坐标为 (5,10),右下角的坐标为 (100,50)。既可以改变矩形轮廓的颜色,也可以改变矩形在光标下的颜色。

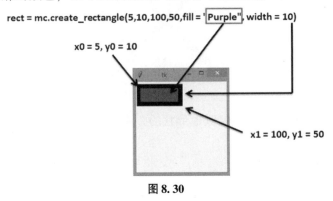

图 8.30

代码:

```
#导入 tkinter
from tkinter import *

#创建窗口实例
mw = Tk()

#创建 Canvas 类实例
mc = Canvas(mw, width = 200, height = 200)

#绘制矩形
rect = mc.create_rectangle(5,10,100,50,fill = "Purple",width = 5,outline
= "Red",activefill = "yellow")

#在窗口对象 mw 上添加 Canvas 对象 mc
mc.pack()

#调用 mainloop()方法
mw.mainloop()
```

输出结果如图 8.31 所示。

图 8.31

将光标放在矩形上时,会注意到颜色变化,如图 8.32 所示。

图 8.32

2. 弧

刚刚学习了如何在 Canvas 上绘制矩形。下面学习如何绘制弧。

示例 8.10

编写代码绘制弧。

在给定坐标定义的矩形空间中绘制弧。create_arc( ) 是创建弧函数。先看一段简单的代码。

```
#导入 tkinter
from tkinter import *

#创建窗口实例
mw = Tk()

#创建 Canvas 类实例
mc = Canvas(mw, width = 200, height = 200)

#绘制弧
arc = mc.create_arc(50,25,150,100)

#在窗口对象 mw 上添加 Canvas 对象 mc
mc.pack()

#调用 mainloop()方法
mw.mainloop()
```

输出结果如图 8.33 所示。

图 8.33

观察如何创建弧，先画矩形轮廓，再观察弧如何放置。
代码：

```
#导入 tkinter
from tkinter import *

#创建窗口实例
mw = Tk()

#创建 Canvas 类实例
mc = Canvas(mw, width = 200, height = 200)

#绘制矩形轮廓
rect = mc.create_rectangle(50,25,150,100,outline = "Red")

#绘制弧
arc = mc.create_arc(50,25,150,100)

#在窗口对象 mw 上添加 Canvas 对象 mc
mc.pack()

#调用 mainloop()方法
mw.mainloop()
```

输出结果如图 8.34 所示。

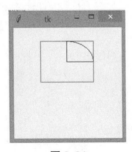

图 8.34

在由坐标（50，25）和（150，100）定义的矩形内创建弧。由于未定义角度，因此默认角度 90°。

按以下方式绘制弧的起点和终点角度：

```
#导入 tkinter
from tkinter import *

#创建窗口实例
mw = Tk()

#创建 Canvas 类实例
mc = Canvas(mw, width = 200, height = 200)
```

```
#绘制弧
arc = mc.create_arc(50,25,150,100, start = 120, extent = 90)

#在窗口对象mw上添加Canvas对象mc
mc.pack()

#调用mainloop()方法
mw.mainloop()
```

输出结果如图 8.35 所示。

图 8.35

为了理解这一点,再观察如何将弧放置在矩形内。以下代码继续创建弧的矩形。

代码:

```
#导入tkinter
from tkinter import *

#创建窗口实例
mw = Tk()

#创建Canvas类实例
mc = Canvas(mw, width = 200, height = 200)

#绘制矩形轮廓
rect = mc.create_rectangle(50,25,150,100,outline = "Red")

#绘制弧
arc = mc.create_arc(50,25,150,100, start = 120, extent = 90)

#在窗口对应mw上添加Canvas对象mc
mc.pack()

#调用mainloop()方法
mw.mainloop()
```

输出结果如图 8.36 所示。

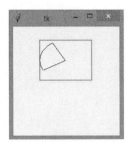

图 8.36

图 8.37 演示了角度的工作原理。同样，要画一个完整的椭圆，create_arc( )函数中的参数 start 可以是 0，extent 可以是 360。下面在 Canvas 上创建椭圆和多边形。

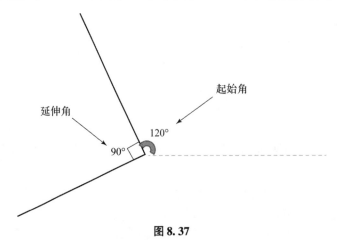

图 8.37

代码：

```
#导入 tkinter
from tkinter import *

#创建窗口实例
mw = Tk()

#创建 Canvas 类实例
mc = Canvas(mw, width = 200, height = 200)

#绘制弧
arc = mc.create_arc(50,25,150,100,start =0, extent = 359)

#在窗口对象 mw 上添加 Canvas 对象 mc
mc.pack()

#调用 mainloop()方法
mw.mainloop()
```

输出结果如图 8.38 所示。

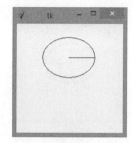

图 8.38

与直线和矩形一样，也能为弧设置特性。

代码：

```
#导入 tkinter
from tkinter import *

#创建窗口实例
mw = Tk()

#创建 Canvas 类实例
mc = Canvas(mw, width = 200, height = 200)

#绘制弧
arc = mc.create_arc(50,25,150,100,start =0, extent = 359,width = 4, fill
= "green", outline = "Purple", activefill = "White")

#在窗口对象 mw 上添加 Canvas 对象 mc
mc.pack()

#调用 mainloop()方法
mw.mainloop()
```

输出结果如图 8.39 所示。

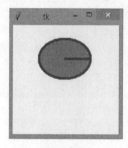

图 8.39

弧的轮廓宽度为 4，颜色为紫色，填充为绿色。当光标放在弧的顶部时，内部颜色变为白色。

输出结果如图 8.40 所示。

图 8.40

示例 8.11

在窗口中创建名为"Create Arc"的按钮，单击该按钮时，提示用户输入弧的参数，并相应显示弧。

代码：

```
#导入 tkinter
from tkinter import *

# 创建按钮函数
def arc_display():

#创建 Canvas 类实例
    mc = Canvas(mw, width = 350, height = 350)

#获取所有值的输入
    x0 = int(input("please enter the value for   x0 : "))
    y0 = int(input("please enter the value for   y0 : "))

    x1 = int(input("please enter the value for   x1 : "))
    y1 = int(input("please enter the value for   y1 : "))

    s_angle = int(input("please enter the value for   start angle : "))
    e_angle = int(input("please enter the value for   extent angle : "))

#使用给定值创建圆弧
    mc.create_arc(x0,y0,x1,y1,start = s_angle,extent = e_angle)

#使 Canvas 在主窗口上可见
    mc.pack()

#创建窗口实例
mw = Tk()

#创建按钮实例
my_first_button = Button(mw, text = "Create Arc", command = arc_display)
```

```
#将要显示的按钮添加到窗口
my_first_button.pack()

#调用 mainloop() 方法
mw.mainloop()
```

输出结果如图 8.41 所示。

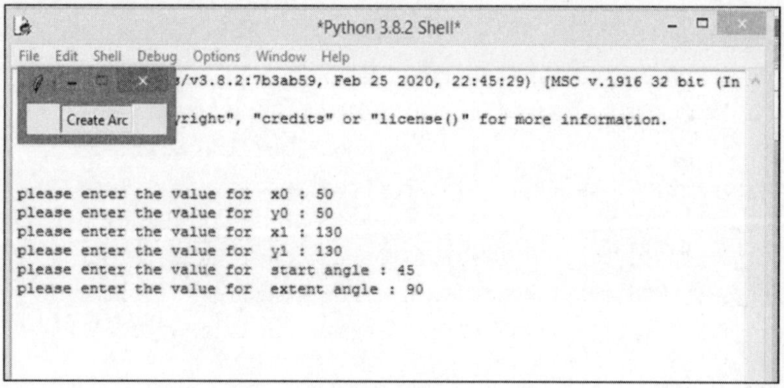

图 8.41

最终结果如图 8.42 所示。

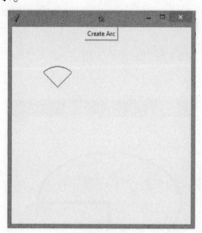

图 8.42

示例 8.12
绘制被吃掉 1/4 比萨的图。
代码:

```
#导入 tkinter
from tkinter import *

#创建窗口实例
mw = Tk()
```

```
#创建 Canvas 类实例
mc = Canvas(mw, width = 350, height = 350)

#绘制多边形
mc.create_arc(30,65,300,250,start = 0, extent = 270, fill = "yellow",width = 5,outline = "black")

#绘制比萨内部的点
mc.create_oval(80,120,94,134,fill = "red",outline = "black")
mc.create_oval(210,80,224,94,fill = "red",outline = "black")
mc.create_oval(110,130,124,144,fill = "red",outline = "black")
mc.create_oval(200,110,214,124,fill = "red",outline = "black")
mc.create_oval(150,70,164,84,fill = "red",outline = "black")
mc.create_oval(150,100,164,114,fill = "red",outline = "black")
mc.create_oval(130,200,144,214,fill = "red",outline = "black")
mc.create_oval(100,200,114,214,fill = "red",outline = "black")

#使 Canvas 在主窗口上可见
mc.pack()

#调用 mainloop()方法
mw.mainloop()
```

输出结果如图 8.43 所示。

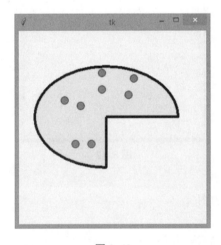

图 8.43

示例 8.13
编写程序绘制椭圆。

代码：

```
#导入 tkinter
from tkinter import *

#创建窗口实例
mw = Tk()

#创建 Canvas 类实例
mc = Canvas(mw, width = 350, height = 350)

#绘制椭圆
mc.create_oval(20,20,200,200,fill = "yellow")

#使 Canvas 在主窗口上可见
mc.pack()

#调用 mainloop()方法
mw.mainloop()
```

输出结果如图 8.44 所示。

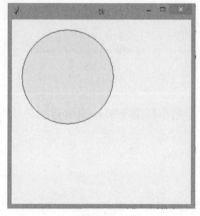

图 8.44

示例 8.14

编写程序绘制多边形。

代码：

```
#导入 tkinter
from tkinter import *

#创建窗口实例
mw = Tk()

#创建 Canvas 类实例
mc = Canvas(mw, width = 350, height = 350)
```

```
#绘制多边形
mc.create_polygon(250,30,200,50,230,90,60,300,100,20, fill = "pink")

#使 Canvas 在主窗口上可见
mc.pack()

#调用 mainloop()方法
mw.mainloop()
```

输出结果如图 8.45 所示。

图 8.45

说明：

如何创建多边形？通过绘制小圆圈来表示这些点。

绘制具有以下点集的多边形。

（1）250，30；

（2）200，50；

（3）230，90；

（4）60，300；

（5）100，20。

在每个点的周围，使用表 8.3 创建小圆圈。

表 8.3

Point	X0(x-1)	Y0(y-1)	X1(x+1)	Y1(y+1)
250, 30	249	29	251	31
200, 50	199	49	201	51
230, 90	229	89	231	91
60, 300	59	299	61	301
100, 20	99	19	101	21

输出结果如图 8.46 所示。

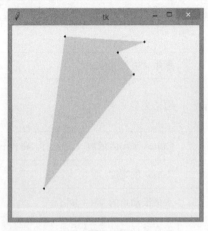

图 8.46

最后看一下 Canvas 的选项。创建 Canvas 的基本代码如下所示。

```
#导入 tkinter
from tkinter import *

#创建窗口实例
mw = Tk()

#创建 Canvas 类实例
mc = Canvas(mw, width = 200, height = 200)

#在窗口对象 mw 上添加 Canvas 对象 mc
mc.pack()

#调用 mainloop()方法
mw.mainloop()
```

表 8.4 描述了 Canvas 控件的可用选项。

表 8.4

选项	描述
bd（默认值为 2）	边框宽度（以像素为单位）
bg	正常背景颜色
confine（默认值为 True）	Canvas 不能滚动到滚动区域外
Cursor	设置 Canvas 使用哪种类型（箭头、圆形、点等）

续表

选项	描述
Height	Canvas 的高度
Highlightcolor	聚焦时突出显示颜色
relief（SUNKEN, RAISED, GROOVE, RIDGE）	边框类型
Scrollregion	Canvas 滚动区域由（左、上、右和下）位置的元组定义
Width	Canvas 的宽度
xscrollcommand	水平滚动条的 set( )方法
yscrollcommand	垂直滚动条的 set( )方法

在 Canvas 中显示文本。

代码：

```
#导入 tkinter
from tkinter import *

#创建窗口实例
mw = Tk()

#创建 Canvas 类实例
mc = Canvas(mw,bg = "pink",width = 400, height = 400)

#定义 Canvas 上的文本字体
fnt = ('Times',15,"bold")

#定义文本
txt = mc.create_text(150,50, text = "Black and White", font = fnt,fill
= "black",activefill = "white")

#在窗口对象 mw 上添加 Canvas 对象 mc
mc.pack()

#调用 mainloop()方法
mw.mainloop()
```

输出结果如图 8.47 所示。

将光标放在顶部时，文本颜色会变为白色。输出结果如图 8.48 所示。

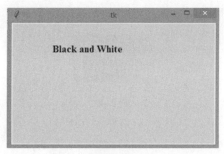

图 8.47

图 8.48

8.5 Frame

8.4 节介绍了 Canvas。本节介绍 Frame。Frame 是容器，用于显示按钮、复选框等控件，有助于组织 GUI。

1. Frame 的用法

代码：

```
#导入语句
from tkinter import *
mw = Tk()

#创建 Frame 对象
mf = Frame(mw)

#将 Frame 附加到主窗口
mf.pack()

#第 1 个按钮
button1 = Button(mf, text = "Left",bg = "red", bd = 10)

#将第 1 个按钮附加到 Frame 的左侧
button1.pack(side = "left")

#第 2 个按钮
button2 = Button(mf, text = "Right",bg = "Green",bd = 10)

#将第 2 个按钮附加到 Frame 的右侧
button2.pack(side = "right")

#调用 mainloop()方法
mw.mainloop()
```

输出结果如图 8.49 所示。

图 8.49

2. Frame 选项

代码：

```
#导入语句
from tkinter import *

mw = Tk()
mw.geometry("200x200")

#创建 Frame 对象
mf = Frame(mw,width =150, height = 150,bg = "green",bd = 20, relief = "sunken")

#将 Frame 附加到主窗口
mf.pack()

#调用 mainloop()方法
mw.mainloop()
```

输出结果如图 8.50 所示。

图 8.50

8.6 标签

与 Canvas 一样，标签也用于显示文本、图像或文字，并能随时更改。相比于 Canvas 控件，标签更易于使用。语法格式如下所示。

```
Label(master, option1,option2…)
```

与其他控件相同，标签也用以逗号分隔的主窗口和选项作为参数来创建实例。

代码：

```python
#导入tkinter
from tkinter import *

#创建主窗口
mw = Tk()

#定义主窗口的尺寸
mw.geometry("200x200")

#创建标签并定义font、width、height、bg、fg等选项
lbl = Label(mw,text = "I am your first label",fg = "blue",bg = "green",
font = ("Comic Sans MS",14,"bold italic"),width = 200, height = 200)

#将标签与主窗口关联
lbl.pack()

#调用mainloop()方法
mw.mainloop()
```

输出结果如图 8.51 所示。

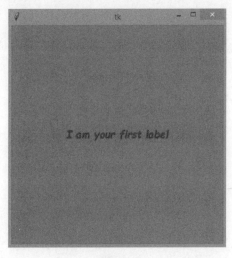

图 8.51

8.7 小项目——秒表

以下代码显示两个按钮。单击"Start"按钮，计时器开始计时；单击"Stop"按钮，显示计时时间。

代码:

```python
#导入语句
from tkinter import *
import time

start_time = 0
end_time = 0
total_time = 0

def time_display(seconds):
#用秒的值除以60 得到分钟的下限值
    minutes = seconds//%60
#用分钟的值除以60 得到小时的下限值
    hours = minutes//60
    minutes = minutes%60
    seconds = seconds%60
    msg = "Time Lapsed = {0}:{1}:{2}".format(int(hours),int(minutes),int(seconds))
    lbl = Label(mw,text = msg,fg = "blue",bg = "white",font =("Comic Sans MS",14,"bold italic"),width = 100, height = 100 )
    lbl.pack(side = "top")

def timer_start():
    global start_time
    print("in start time")
    start_time = time.time()
    print(start_time)

def timer_end():
    end_time = time.time()
    total_time = end_time - start_time
    time_display(int(total_time))

#创建主窗口
mw = Tk()
mw.geometry("300 x300")
#Create a frame Object
mf = Frame(mw)

#将Frame 附加到主窗口
mf.pack(side = "bottom")

#第1 个按钮
button1 = Button(mf, text = "Start",bg = "green", bd = 10,command = timer_start)

#将第1 个按钮附加到Frame 的左侧
button1.pack(side = "left")
```

```
#第 2 个按钮
button2 = Button(mf,text = "Stop",bg = "red",bd = 10,command = timer_end)

#将第 2 个按钮附加到 Frame 的右侧
button2.pack(side = "right")

#调用 mainloop()方法
mw.mainloop()
```

输出结果如图 8.52 所示。

图 8.52

单击"Start"按钮,持续一段时间后单击"Stop"按钮。输出结果如图 8.53 所示。

图 8.53

8.8 列表框

列表框用于显示项目列表。创建列表框的语法格式如下所示。

Listbox(master,options…)

代码:

```
from tkinter import *
mw = Tk()
mw.geometry("300×300")
lstbx = Listbox(mw,bg = "skyblue",bd = 10,relief = "raised", font = ("Times",
"15", "bold italic"),fg = "red",height = 15,width = 10, highlightcolor = "pink",
highlightthickness = 10,selectbackground = "blue",selectmode = "single")
lstbx.insert(1,"Monday")
lstbx.insert(2,"Tuesday")
lstbx.insert(3,"Wednesday")
lstbx.insert(4,"Thursday")
lstbx.insert(5,"Friday")
lstbx.insert(6,"Saturday")
lstbx.insert(7,"Sunday")
lstbx.pack()
mw.mainloop()
```

输出结果如图 8.54 所示。

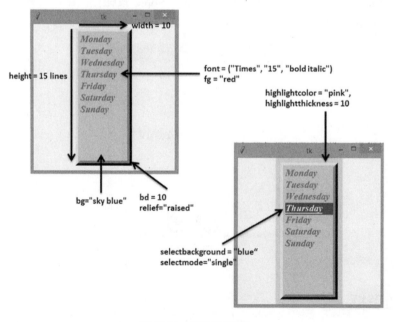

图 8.54　列表框选项

列表框的主要方法如表 8.5 所示。

表 8.5

方法	说明
activate(index)	根据参数传递值选择指定的行
curselection( )	返回所选元素的行号或打包在元组中的元素（从 0 开始计数）。如果未选择任何内容，则返回空元组
delete(first, last = None)	删除参数传递的行范围。如果没有值传递第 2 个参数，则删除具有第 1 个参数值的单行
get(first, last = None)	检索参数指定范围内的行文本。如果只给出 1 个参数，则返回该行文本
index(i)	定位列表框的可见部分，让包含索引的行位于顶部
insert(index, * elements)	在 index 指定行前将一行或多行插入列表框
nearest(y)	返回相对于列表框控件最接近 y 坐标的行索引
see(index)	调整列表框位置，使索引行可见
size( )	返回列表框的大小或行数
xview( )	列表框水平滚动
yview( )	列表框垂直滚动

8.9 菜单按钮和菜单

本节介绍如何创建菜单按钮。

步骤 1：创建主窗口。

代码：

```
#导入 tkinter
from tkinter import *

#创建主窗口( +120 +120 是与 xy 的距离)
mw = Tk()
mw.geometry("300×300 +100 +100") #解释如下

#调用 mainloop()方法
mw.mainloop()
```

创建的窗口如图 8.55 所示。

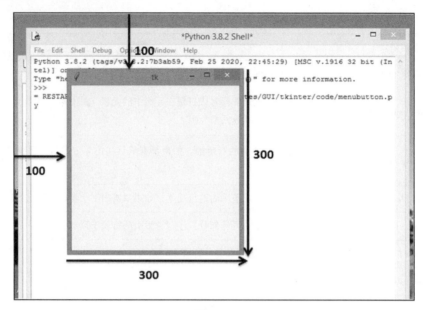

图 8.55

查看代码,发现以下语句有问题:

```
mw.geometry("300 x300 +100 +100")
```

其中,300×300 表示窗口大小;+100+100 表示窗口的屏幕显示位置。

步骤 2:创建菜单按钮。

代码:

```
#创建菜单按钮
mb = Menubutton(mw,text = "File")
mb.grid()
```

相比于其他控件,创建菜单按钮时传递的第 1 个参数是主窗口,示例中是 mw;第 2 个参数是菜单按钮显示的文本。

此代码放在 mainloop( ) 方法之前。

代码:

```
#导入 tkinter
from tkinter import *

#创建主窗口
mw = Tk()
mw.geometry("300 x300 +100 +100")
```

```
#创建菜单按钮
mb = Menubutton(mw,text = "File")
mb.grid()

#调用mainloop()方法
mw.mainloop()
```

输出结果如图 8.56 所示。

图 8.56

步骤 3：在菜单按钮中创建下拉菜单。

将以下代码行添加到步骤 4 的代码中。Menu( )函数的第 1 个参数 mb 是菜单按钮，第 2 个参数 tearoff = 0。菜单默认支持 tearoff，也就是通过将其与主菜单分离来创建浮动菜单。因此，如果要禁用此功能，就将其设置为 0。不了解此功能的人只需使用此设置，阅读示例末尾的说明。代码如下所示。

```
#创建下拉菜单
mb.menu = Menu(mb,tearoff = 0)
mb["menu"] = mb.menu
```

因此，到现在为止开发的代码应该是这样的：

```
#导入 tkinter
from tkinter import *

#创建主窗口( +120 +120 是与 xy 的距离)
mw = Tk()
mw.geometry("300 x300 +100 +100")

#创建菜单按钮
mb = Menubutton(mw,text = "File")
mb.grid()
```

```
#创建下拉菜单
mb.menu = Menu(mb,tearoff=0)
mb["menu"] = mb.menu

#调用mainloop()方法
mw.mainloop()
```

步骤4：添加下拉菜单命令。代码如下所示。

```
#向下拉菜单中添加命令
mb.menu.add_command(label = "New File")
mb.menu.add_command(label = "Open")
mb.menu.add_command(label = "Recent Files")
mb.menu.add_command(label = "Save")
mb.menu.add_command(label = "Save As")
mb.menu.add_command(label = "Print")
mb.menu.add_command(label = "Close")
mb.menu.add_command(label = "Exit")
```

创建带有下拉菜单的菜单按钮的代码如下所示。

```
#导入tkinter
from tkinter import *

#创建主窗口( +120 +120 是与xy的距离)
mw = Tk()
mw.geometry("300x300 +100 +100")

#创建菜单按钮
mb = Menubutton(mw,text = "File")
mb.grid()

#创建下拉菜单
mb.menu = Menu(mb,tearoff=0)
mb["menu"] = mb.menu

#向下拉菜单中添加命令
mb.menu.add_command(label = "New File")
mb.menu.add_command(label = "Open")
mb.menu.add_command(label = "Recent Files")
mb.menu.add_command(label = "Save")
mb.menu.add_command(label = "Save As")
mb.menu.add_command(label = "Print")
mb.menu.add_command(label = "Close")
mb.menu.add_command(label = "Exit")

#调用mainloop()方法
mw.mainloop()
```

输出结果如图 8.57 所示。

图 8.57

在两个命令之间添加分隔符。

代码：

```
mb.menu.add_command(label = "New File")
mb.menu.add_command(label = "Open")
mb.menu.add_command(label = "Recent Files")
mb.menu.add_separator()
mb.menu.add_command(label = "Save")
mb.menu.add_command(label = "Save As")
mb.menu.add_command(label = "Print")
mb.menu.add_separator()
mb.menu.add_command(label = "Close")
mb.menu.add_command(label = "Exit")
```

输出结果如图 8.58 所示。

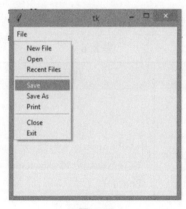

图 8.58

注意步骤 3 中提到的 tearoff 功能。

tkinter 菜单默认支持 tearoff，就是通过将其与主菜单分离来创建浮动菜单。因此，要禁用此功能,将其设置为0。

转到创建下拉菜单的代码。

```
#创建下拉菜单
mb.menu = Menu(mb,tearoff=0)
mb["menu"] = mb.menu
```

下面去除tearoff设置。代码如下所示。

```
#创建下拉菜单
mb.menu = Menu(mb)
mb["menu"] = mb.menu
```

执行代码,发现在菜单选项的顶部有虚线。单击虚线,菜单将与主菜单分离输出,如图8.59所示。

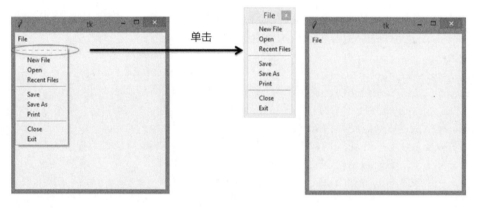

图 8.59

表8.6所示为菜单按钮的主要选项。

表8.6

选项	说明	代码	输出结果
activeback-ground	鼠标放在菜单上时的背景颜色	Menubutton (mw, text = "File",activebackground = "Red")	activebackground = "Red"

续表

选项	说明	代码	输出结果
activefore-ground	鼠标放在菜单上时的前景色	`Menubutton ( mw, text = "File",activeforeground = "Red")`	activeforeground = "Red"
anchor	文本位置选项：n、ne、e、se、s、sw、w、nw 或 center。center 是默认值	`Menubutton ( mw, text = "File",height = 5, width = 5,anchor = "se")`	anchor = "se"
bg	标签后面显示的背景颜色	`Menubutton ( mw, text = "File",bg = "red")`	bg = "red"
bitmap	可用的标准位图：error、gray75、gray50、gray25、gray12、hourglass、info、questhead、question 和 warning，也可添加自定义位图	`Menubutton ( mw, text = "File",bitmap = "questhead")`	bitmap = "questhead"
direction	菜单显示方向。其值为 left、right 或 above	`mb = Menubutton(mw,text = "File", direction = "left")`	direction = "left"

选项	说明	代码	输出结果
fg	背景色	`Menubutton(mw,text = "File",fg = "red")`	fg = "red"
image	菜单按钮的显示图像	`img = PhotoImage(file = "f:\\my_images\\thumb2.gif")` `mb = Menubutton(mw,text = "File",image = img)`	image = img
padx, pady	padx 表示文本左侧和右侧空间；pady 表示文本上方和下方空间	`Menubutton(mw,text = "File",padx =10,pady = 10)`	padx = 10,pady = 10
relief	边形状	`Menubutton(mw,text = "File",relief = "raised")`	relief = "raised"
state	设置为禁用时，禁用按钮	`Menubutton(mw,text = "File",state = "disable")`	state = "disable"

表 8.7 所示为菜单的主要选项。

表 8.7

选项	说明	代码	输出结果
activeback-ground	鼠标下面的颜色	`Menu(mb,tearoff = 0, activebackground = "green")`	activebackground = "green"

续表

选项	说明	代码	输出结果
activeborder-width	鼠标下方围绕选项绘制的边框宽度	`Menu (mb, tearoff = 0, activeborder-width = 15)`	**activeborderwidth = 15**
activefore-ground	鼠标下方的文本颜色	`mb.menu = Menu ( mb, tearoff = 0, active-foreground = "red")`	**activeforeground = "red"**
bg	背景颜色	`mb.menu = Menu ( mb, tearoff =0,bg = "red")`	**bg = "red"**
disabledfore-ground	禁用项目的文本颜色	—	—
Postcommand	每次调出此菜单时都会调用程序	—	—

8.10 单选按钮

单选按钮是多选按钮的实现。单选按钮向用户呈现多个对象，并让用户只选择其中一个。

使用 Radiobutton( )方法创建单选按钮。

语法格式如下所示。

```
Radiobutton(root,options)
```

代码：

```
#导入 tkinter
from tkinter import *
```

```
#创建主窗口
mw = Tk()
mw.geometry("300x300")

#创建单选按钮
rb1 = Radiobutton(mw,text = "C")
rb2 = Radiobutton(mw,text = "C++")
rb3 = Radiobutton(mw,text = "Java")
rb4 = Radiobutton(mw,text = "Python")
rb5 = Radiobutton(mw,text = "Perl")

rb1.grid(row = 0,column = 0,sticky = W)
rb2.grid(row = 1,column = 0,sticky = W)
rb3.grid(row = 2,column = 0,sticky = W)
rb4.grid(row = 3,column = 0,sticky = W)
rb5.grid(row = 4,column = 0,sticky = W)

#调用mainloop()方法
mw.mainloop()
```

输出结果如图8.60所示。

图 8.60

或者使用如下代码：

```
#导入tkinter
from tkinter import *

#创建主窗口
mw = Tk()
mw.geometry("300x300")

#创建单选按钮
rb1 = Radiobutton(mw,text = "C")
rb2 = Radiobutton(mw,text = "C++")
rb3 = Radiobutton(mw,text = "Java")
```

```
rb4 = Radiobutton(mw,text = "Python")
rb5 = Radiobutton(mw,text = "Perl")

rb1.pack(anchor = W)
rb2.pack(anchor = W)
rb3.pack(anchor = W)
rb4.pack(anchor = W)
rb5.pack(anchor = W)

#调用mainloop()方法
mw.mainloop()
```

输出结果如图 8.61 所示。

图 8.61

8.11 滚动条和滑块

滚动条用于在其他控件中滚动文本。从左到右滚动使用水平滚动条，从上到下滚动使用垂直滚动条。以下分步说明。

步骤 1：创建主体结构。代码如下所示。

```
import tkinter as tk
mw = tk.Tk()
mw.geometry("100×100")
mw.mainloop()
```

输出结果如图 8.62 所示。

图 8.62

步骤 2：创建滚动条并将其关联到主窗口。代码如下所示。

```
#垂直滚动条
sb = tk.Scrollbar(mw)
sb.pack(side = tk.RIGHT,fill = "y")
```

输出结果如图 8.63 所示。

图 8.63

步骤 3：创建文本框并布局到主窗口。代码如下所示。

```
#创建文本框
tb = tk.Text(mw,height = 500,width = 500,yscrollcommand = sb.set,bg = "skyblue")
tb.pack()
```

输出结果如图 8.64 所示。

图 8.64

步骤 4：使用 config( )方法将滚动条配置到文本框中。

config( )方法用于对象初始化后访问对象属性，以便运行时对其进行设置。代码如下所示。

```
#将滚动条配置到文本框
sb.config(command = tb.yview)
```

步骤 5：创建文本框内容。代码如下所示。

```
list_of_month = "January\nFebruary\nMarch\nApril\nMay\nJune\nJuly\\
nAugust\nSeptember\nOctober\nNovember\nDecember"
```

在文本框中插入内容。代码如下所示。

```
#在文本框中插入内容
list_of_month = "January\nFebruary\nMarch\nApril\nMay\nJune\nJul\nAu-
gust\nSeptember\nOctober\nNovember\nDecember"
```

现在代码如图 8.65 所示。

```
import tkinter as tk

#content for textbox
list_of_month = "January\nFebruary\nMarch\nApril\nMay\nJune\nJul\nAugust\nSeptem

mw = tk.Tk()
mw.geometry("100x100")

#Vertical Scrollbar
sb = tk.Scrollbar(mw)
sb.pack(side = tk.RIGHT,fill="y")

#Create a Textbox
tb = tk.Text(mw,height = 500, width = 500, yscrollcommand = sb.set, bg="sky blue
tb.pack()

sb.config(command = tb.yview)
tb.insert(tk.END,list_of_month)

mw.mainloop()
```

图 8.65

输出结果如图 8.66 所示。

图 8.66

步骤 6：添加水平滚动条。代码如下所示。

```
sb2 = tk.Scrollbar(mw,orient = tk.HORIZONTAL)
sb2.pack(side = tk.BOTTOM, fill = "x")
```

步骤 7：添加水平内容。代码如下所示。

```
list_of_days = "Monday Tuesday Wednesday Thursday Friday Saturday Sunday"
tb.insert(tk.END,list_of_days)
```

步骤 8：定义 xscrollcommand。

在步骤 3 中，创建了一个文本框，并设置了 yscrollcommand，代码如下所示。

```
tb = tk.Text(mw,height = 500, width = 500, yscrollcommand = sb.set, bg = "skyblue")
```

在此声明中再添加以下两个事件：

（1） xscrollcommand = sb2.set；
（2） wrap = "none"。

因此，该语句如下所示。

```
tb = tk.Text(mw,height = 500, width = 500, yscrollcommand = sb.set, xscrollcommand = sb2.set, wrap = "none", bg = "skyblue")
```

配置 x 滚动条到文本框。代码如下所示。

```
sb2.config(command = tb.xview)
```

代码：

```
import tkinter as tk

#文本框内容
list_of_month = "January\nFebruary\nMarch\nApril\nMay\nJune\nJul\nAugust\nSeptember\nOctober\nNovember\nDecember\n"
list_of_days = "Monday Tuesday Wednesday Thursday Friday Saturday Sunday"

mw = tk.Tk()
mw.geometry("100x100")

#垂直滚动条
sb = tk.Scrollbar(mw)
sb.pack(side = tk.RIGHT,fill = "y")

#水平滚动条
sb2 = tk.Scrollbar(mw,orient = tk.HORIZONTAL)
sb2.pack(side = tk.BOTTOM, fill = "x")

#创建文本框
tb = tk.Text(mw,height = 500, width = 500, yscrollcommand = sb.set, xscrollcommand = sb2.set, wrap = "none", bg = "skyblue")
tb.pack()

sb.config(command = tb.yview)
sb2.config(command = tb.xview)

tb.insert(tk.END,list_of_month)
tb.insert(tk.END,list_of_days)

mw.mainloop()
```

输出结果如图 8.67 所示。

图 8.67

8.12 文本框

在前面的练习中已处理过文本框。本节介绍该控件的更多内容。文本框提供多行文本区域且灵活方便,可以实现很多任务。通常程序开发人员在需要显示多行区域时使用该控件。另外,文本区域还用于显示链接图像和 HTML 等。在前面的示例中已经学习了创建文本框的基础知识。现在简单回顾一下。

步骤1:基本结构的主要代码如下。

```
from tkinter import *
mw = Tk()
mw.geometry("150x150")
ta = Text(mw)
ta.insert(INSERT,"Sky is the limit!!")
ta.pack()
mw.mainloop()
```

输出结果如图 8.68 所示。

图 8.68

步骤2:在文本框中插入图像。代码如下所示。

```
img = PhotoImage(file = "f:\my_images\sky.gif")
ta.image_create(END,image = img)
```

输出结果如图 8.69 所示。

图 8.69

步骤3:增加滚动条。代码如下所示。

```
#垂直滚动条
sb = Scrollbar(mw)
sb.pack(side = RIGHT,fill = "y")

#水平滚动条
sb2 = Scrollbar(mw,orient = HORIZONTAL)
sb2.pack(side = BOTTOM, fill = "x")

sb.config(command = ta.yview)
sb2.config(command = ta.xview)
```

此外,将语句 ta = Text(mw)更改为:

```
ta = Text(mw,yscrollcommand = sb.set,xscrollcommand = sb2.set)
```

最终代码如下所示。

```
from tkinter import *
mw = Tk()
mw.geometry("150x150")
img = PhotoImage(file = "f:\\my_images\\sky.gif")

#垂直滚动条
sb = Scrollbar(mw)
sb.pack(side = RIGHT,fill = "y")

#水平滚动条
sb2 = Scrollbar(mw,orient = HORIZONTAL)
sb2.pack(side = BOTTOM, fill = "x")

#创建文本框
ta = Text(mw,yscrollcommand = sb.set,xscrollcommand = sb2.set)

#设置文本框的滚动条
sb.config(command = ta.yview)
sb2.config(command = ta.xview)

#在文本框中插入内容
ta.insert(INSERT,"Sky is the limit!!")
ta.image_create(END,image = img)

ta.pack()
mw.mainloop()
```

8.13 Spinbox

Spinbox 是标准 tkinter 控件。

代码：

```
from tkinter import *
mw = Tk()
mw = Spinbox(mw)
mw.pack()
mainloop()
```

输出结果如图 8.70 所示。

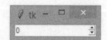

图 8.70

知识要点

- Python 提供 tkinter 模块用于创建 GUI。
- tkinter 是使用工具命令语言（TCL）的 TK 模块类。
- TK 表示工具包，TCL 使用其创建图形。
- TK 提供标准 GUI，可供 Python 等动态编程语言使用。
- Python 程序开发人员使用 tkinter 模块访问 TK。
- Python 提供多种开发 GUI 的方法。
- Python 和 tkinter 是创建 GUI 的最简单的方法。
- tkinter 是最常用的 GUI 模块。
- mainloop() 方法是无限循环，用于运行应用程序。
- 只要窗口未关闭，就会等待事件发生并响应该事件。
- title() 方法用于设置窗口标题。
- geometry() 函数用于更改窗口大小。
- wm_iconbitmap() 方法用于实现新图替换旧图，图只能是 ".ico" 格式。
- tkinter 模块为 GUI 应用程序提供了多种类型的控件，如标签、按钮、单选按钮等。
- 使用控件的步骤。
    - 步骤 1：导入 tkinter。
    - 步骤 2：创建主窗口。
    - 步骤 3：编写控件响应代码。
    - 步骤 4：调用 mainloop() 方法。

简答题

1. 使用 GUI 应用程序的第一步，也是最重要的一步是什么？
答：导入 tkinter。
2. tkinter 是 Python 开发 GUI 的唯一途径吗？
答：不是。
3. tkinter 是否是提供 Python 创建 GUI 的最简单快捷的方法？
答：是。
4. Python 中 tkinter 模块的 mainloop() 方法的作用是什么？
答：保持窗口。
5. 准备运行应用程序时，必须调用哪个方法？
答：mainloop()。
6. 匹配表 8.8 中的内容。

表 8.8

创建主窗口	title()
设置使用窗口的标题	.ico
将图像绑定到主窗口的方法	geometry()
图像文件	Tk()
改变窗口大小	mainloop()
运行 GUI 应用的无限循环	wm_iconbitmap()

答：匹配后的内容如表 8.9 所示。

表 8.9

创建主窗口	Tk()
设置使用窗口的标题	title()
将图像绑定到主窗口的方法	wm_iconbitmap()
图像文件	.ico
改变窗口大小	geometry()
运行 GUI 应用的无限循环	mainloop()

第 9 章

MySQL和Python图形用户界面

> 前面已经介绍了使用 Python 编程访问 MySQL 数据库的方法。本章介绍如何创建功能齐全的应用程序，以帮助用户使用图形用户界面直接访问数据库。

知识结构
- MySQLdb 数据库
- 使用 GUI 创建表
- 使用 GUI 插入数据
- 创建 GUI 以检索结果

目标
完成本章的学习后，读者应能够编写 Python 函数代码，以便从图形用户界面更新数据库。

9.1 MySQLdb 数据库

为了使 Python 程序与数据库交互，首先要连接到数据库服务器（本示例中为 MySQL）。为连接到 MySQL 服务器，需提供以下信息。

（1）host：MySQL 服务器的 IP 地址，本示例中是 localhost。
（2）user：连接到数据库的用户名。
（3）passwd：连接数据库的密码。
（4）database：Python 程序连接数据库的名称。

语法格式如下所示。

```
Your_connection = msql.connect(host = host_name,user = user_name,passwd = database_password,charset ='utf8', database = database_name)
```

注意：对于更高版本的 MySQL，不需要提供 charset='utf8' 参数，这是 MySQL 6.0 的要求。访问 SQL Server 的代码如下所示。在本示例中，数据库名称为 textile；用户名为 shopkeeper；密码为 shoptoday。

```
import mysql.connector as msql
pycon = msql.connect(host = 'localhost', user = 'shopkeeper', passwd = 'shop-
today', database = 'textile', charset = 'utf8')
---- your code comes here ----
pycon.close()
```

连接 MySQL 服务器的更快的方法是存储所有必要信息。

配置文件连接到服务器，然后连接数据库。创建字典存储连接所需的全部信息，代码如下所示。

```
dbConfig = {'user':'shopkeeper','password':'shoptoday','host':'local-
host','database':'textile','charset':'utf8'}
```

在创建数据库连接时，只需解压字典中的存储值。代码如下所示。

```
pycon = msql.connect(**dbConfig)
```

因此，连接数据库的基本代码如下所示。

```
import mysql.connector as msql
dbConfig = {'user':'shopkeeper','password':'shoptoday','host':'local-
host','database':'textile','charset':'utf8'}
pycon = msql.connect(**dbConfig)
---- your code comes here ---
pycon.close()
```

两段代码给出了相同的结果。但是，要知道简单地对登录信息进行硬编码并不安全，尤其在处理 Web 应用程序时，这虽然是最快捷的方法，但绝对不安全。在理想情况下，应将所有登录信息存入字典并将其保存在 Python 文件中。

因此，创建 Python 文件 "credential.py" 并保存在当前工作目录中。
粘贴以下代码：

```
dbConfig = {'user':'shopkeeper','password':'shoptoday','host':'local-
host','database':'textile','charset':'utf8'}
```

运行以下代码：

```
import mysql.connector as msql
import credentials as c
pycon = msql.connect(**c.dbConfig)
print(pycon)
pycon.close()
```

输出：

```
<mysql.connector.connection.MySQLConnection object at 0x037035F8>
```

注意：在第 4 章中已经学习了如何使用 connect() 方法连接 MySQL，该方法返回连接对象。如果连接成功，调用函数 print(connection_object)，连接对象的信息会显示如下。

因此，将登录信息保存在字典对象单独文件中并将其解压到代码中属于安全连接。

现在，创建游标对象并执行 SHOW DATABASES 命令。

代码：

```python
import mysql.connector as msql
import credentials as c
pycon = msql.connect(**c.dbConfig)
mycursor = pycon.cursor()
mycursor.execute('SHOW DATABASES;')
result_set = mycursor.fetchall()
for result in result_set:
    print(result)
pycon.close()
print('Done!! ')
```

输出：

```
('information_schema',)
('textile',)
Done!!
```

对于数据库与 GUI 的交互，下一节将做说明。

注意：关于游标对象的更多信息，请参阅第 4 章内容。

9.2 使用 GUI 创建表

尝试创建带有 GUI 的表。这个示例非常简单，但仍有改进空间。在理想情况下，最好使用简单的方法了解事物的运作方式。

下面在 textile 数据库中创建 Fabric 表。代码如下所示。

```sql
CREATE TABLE FABRIC(
    FABRIC_ID SMALLINT NOT NULL AUTO_INCREMENT,
    FABRIC_NAME CHAR(20) NOT NULL,
    IMPORT_EXPORT CHAR(20) NOT NULL,
    COST_PER_METER SMALLINT NOT NULL,
    COLORS_AVAILABLE SMALLINT NOT NULL,
    PRIMARY KEY (FABRIC_ID)
);
```

要创建和使用的 GUI 如图 9.1 所示，需要表名、每列详细信息和主键。加入此信息并将语句传递给数据库。

输出结果如图 9.1 所示。

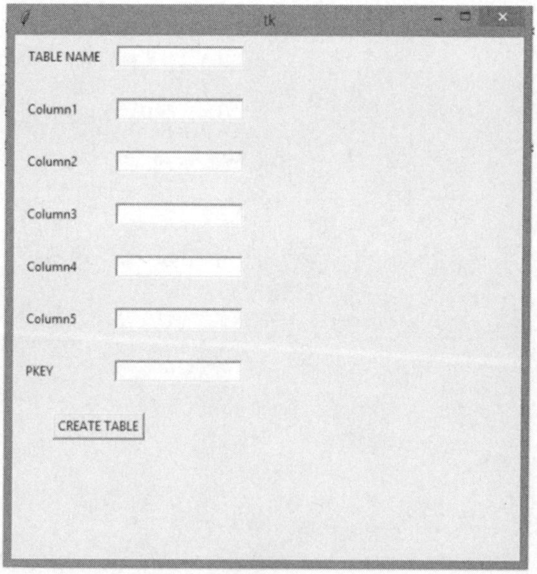

图 9.1

步骤 1：准备 GUI。

准备 GUI 创建表。代码如下所示。

```python
#导入 tkinter
from tkinter import *

#创建窗口实例
mw = Tk()
mw.geometry('500x500')
table_name = Label(mw, text = 'TABLE NAME')
table_name.place(x =10, y =10)

table_name_E = Entry(mw,bd = 2)
table_name_E.place(x = 100, y =10)

column1 = Label(mw, text = 'Column1')
column1.place(x = 10 , y = 60)

column1_E = Entry(mw,bd = 2)
column1_E.place(x = 100 , y = 60 )

column2 = Label(mw, text = 'Column2')
column2.place(x = 10 , y = 110)

column2_E = Entry(mw,bd = 2)
column2_E.place(x = 100 , y = 110 )
```

```
column3 = Label(mw, text = 'Column3')
column3.place(x = 10, y = 160 )

column3_E = Entry(mw,bd = 2)
column3_E.place(x = 100, y = 160 )

column4 = Label(mw, text = 'Column4')
column4.place(x = 10, y = 210)

column4_E = Entry(mw,bd = 2)
column4_E.place(x = 100, y = 210 )

column5 = Label(mw, text = 'Column5')
column5.place(x = 10, y = 260)

column5_E = Entry(mw,bd = 2)
column5_E.place(x = 100 , y = 260 )

columnpk = Label(mw, text = 'PKEY')
columnpk.place(x = 10, y = 310)

columnpk_E = Entry(mw,bd = 2)
columnpk_E.place(x = 100 , y = 310 )

create_button = Button(mw,text = 'CREATE TABLE')
create_button.place(x = 40 , y = 360)

mw.mainloop()
```

接下来,单击"CREATE TABLE"按钮时,查询信息传递到数据库。

步骤2:导入语句连接数据库。在创建函数前,导入以下内容:

```
#导入mysql.connector
import mysql.connector as msql

#导入credentials.py,其中包含登录详细信息
import credentials as c
```

步骤3:定义创建表函数。代码如下所示。

```
def create_tab():
    pycon = msql.connect( **c.dbConfig)
    mycursor = pycon.cursor()
    statement = "CREATE TABLE " + str(table_name_E.get()) + "(" + str(column1_E.get()) + ", " + str(column2_E.get()) + ", " + str(column3_E.get()) + ", " + str(column4_E.get()) + ", " + str(column5_E.get()) + ", " + "PRIMARY KEY(" + str(columnpk_E.get()) + "));"
    print("Passing following information to MySQL textile
    database:" + statement)
    mycursor.execute(statement)
    pycon.close()
print('Done!! ')
```

步骤4：单击"CREATE TABLE"按钮，调用create_tab()函数。代码如下所示。

```
create_button = Button(mw,text = 'CREATE TABLE', command = create_tab)
```

代码：

```
#导入tkinter
from tkinter import *

#导入mysql.connector
import mysql.connector as msql

#导入credentials.py,其中包含登录详细信息
import credentials as c

#定义创建表的函数
def create_tab():
    pycon = msql.connect(**c.dbConfig)
    mycursor = pycon.cursor()
    statement = "CREATE TABLE " + str(table_name_E.get()) + "(" + str(column1_E.get()) + ", " + str(column2_E.get()) + ", " + str(column3_E.get()) + ", " + str(column4_E.get()) + ", " + str(column5_E.get()) + ", " + " PRIMARY KEY(" + str(columnpk_E.get()) + "));"
    print("Passing following information to MySQL textile database:" + statement)
    
    mycursor.execute(statement)
    pycon.close()

#创建窗口实例
mw = Tk()
mw.geometry('500x500')

table_name = Label(mw, text = 'TABLE NAME')
table_name.place(x =10, y =10)

table_name_E = Entry(mw,bd = 2)
table_name_E.place(x = 100, y =10)

column1 = Label(mw, text = 'Column1')
column1.place(x = 10 , y = 60)

column1_E = Entry(mw,bd = 2)
column1_E.place(x = 100 , y = 60 )

column2 = Label(mw, text = 'Column2')
column2.place(x = 10 , y = 110)

column2_E = Entry(mw,bd = 2)
column2_E.place(x = 100 , y = 110 )
```

```
    column3 = Label(mw, text = 'Column3')
    column3.place(x = 10, y = 160 )

    column3_E = Entry(mw,bd = 2)
    column3_E.place(x = 100, y = 160 )

    column4 = Label(mw, text = 'Column4')
    column4.place(x = 10, y = 210)

    column4_E = Entry(mw,bd = 2)
    column4_E.place(x = 100 , y = 210 )

    column5 = Label(mw, text = 'Column5')
    column5.place(x = 10, y = 260)

    column5_E = Entry(mw,bd = 2)
    column5_E.place(x = 100 , y = 260 )

    columnpk = Label(mw, text = 'PKEY')
    columnpk.place(x = 10, y = 310)

    columnpk_E = Entry(mw,bd = 2)
    columnpk_E.place(x = 100 , y = 310 )

    create_button = Button(mw,text = 'CREATE TABLE', command = create_tab)
create_button.place(x = 40 , y = 360)

    mw.mainloop()
```

输出结果如图 9.2 所示。

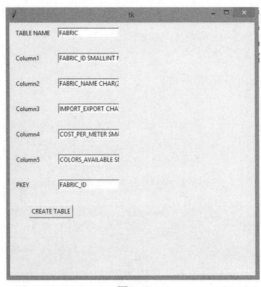

图 9.2

输出：

```
Passing following information to MySQL textile database:CREATE TABLE FABRIC
(FABRIC_ID SMALLINT NOT NULL AUTO_INCREMENT, FABRIC_NAME CHAR(20) NOT NULL,
IMPORT_EXPORT CHAR(20) NOT NULL, COST_PER_METER SMALLINT NOT NULL, COLORS_AVAIL-
ABLE SMALLINT NOT NULL, PRIMARY KEY(FABRIC_ID));
Done!!
```

输出结果如图 9.3 所示。

mysql> SHOW COLUMNS FROM FABRIC;

Field	Type	Null	Key	Default	Extra
FABRIC_ID	smallint(6)	NO	PRI	NULL	auto_increment
FABRIC_NAME	char(20)	NO			
IMPORT_EXPORT	char(20)	NO			
COST_PER_METER	smallint(6)	NO			
COLORS_AVAILABLE	smallint(6)	NO			

5 rows in set (0.08 sec)

图 9.3

9.3　使用 GUI 插入数据

本节介绍如何使用 GUI 插入数据。

步骤 1：GUI 代码。

对 9.2 节中的 GUI 代码进行微小更改，如下所示。

```
#导入 tkinter
from tkinter import *
#创建窗口实例
mw = Tk()
mw.geometry('400x400')
table_name = Label(mw, text = 'TABLE NAME')
table_name.place(x =10, y =10)

table_name_E = Entry(mw,bd = 2)
table_name_E.place(x = 150, y =10)

column1 = Label(mw, text = 'FABRIC_NAME')
column1.place(x = 10 , y = 60)

column1_E = Entry(mw,bd = 2)
column1_E.place(x = 150 , y = 60 )

column2 = Label(mw, text = 'IMPORT_EXPORT')
column2.place(x = 10 , y = 110)

column2_E = Entry(mw,bd = 2)
column2_E.place(x = 150 , y = 110 )
```

```
column3 = Label(mw, text = 'COST_PER_METER')
column3.place(x = 10, y = 160 )

column3_E = Entry(mw,bd = 2)
column3_E.place(x = 150, y = 160 )

column4 = Label(mw, text = 'COLORS_AVAILABLE')
column4.place(x = 10, y = 210)

column4_E = Entry(mw,bd = 2)
column4_E.place(x = 150, y = 210 )

insert_button = Button(mw,text = 'INSERT VALUES')
insert_button.place(x = 40 , y = 260)

mw.mainloop()
```

输出结果如图9.4所示。

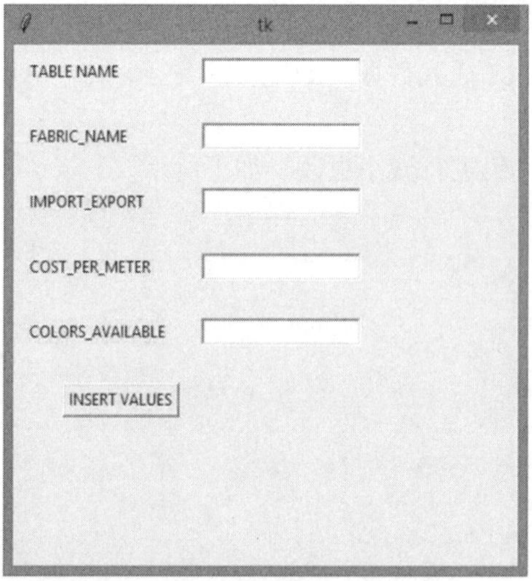

图9.4

步骤2：定义函数。代码如下所示。

```
#导入mysql.connector
import mysql.connector as msql
#导入credentials.py,其中包含登录详细信息
import credentials as c

#定义创建表的函数
def insert_val():
    pycon = msql.connect( **c.dbConfig)
```

```
        mycursor = pycon.cursor()
        statement = "INSERT INTO " +str(table_name_E.get()) + " (FABRIC_
        NAME,IMPORT_EXPORT,COST_PER_METER,COLORS_AVAILABLE) VAlUES
        (" +"'" +str(column1_E.get()) +"'," +"'" +str(column2_E.get()) +"',
        " +" {0},{1});".format(int(column3_E.get()),int(column4_E.get()))
        print("Passing following information to MySQL textile database:"
         +statement)
        mycursor.execute(statement)
        pycon.commit() pycon.close()
print('Done!!')
```

步骤3：使用"INSERT VALUES"按钮映射函数。代码如下所示。

```
insert_button = Button(mw,text = 'INSERT VALUES', command = insert_val)
insert_button.place(x = 40 , y = 260)
```

代码：

```
#导入tkinter
from tkinter import *
#导入mysql.connector
import mysql.connector as msql
#导入credentials.py,其中包含登录详细信息
import credentials as c

#定义创建表的函数
def insert_val():
    pycon = msql.connect(**c.dbConfig)
    mycursor = pycon.cursor()
    statement = "INSERT INTO " +str(table_name_E.get()) + "(FABRIC_
    NAME,IMPORT_EXPORT,COST_PER_METER,COLORS_AVAILABLE) VAlUES
    (" +"'" +str(column1_E.get()) +"'," +"'" +str(column2_E.get()) +"',
    " +" {0},{1});".format(int(column3_E.get()),int(column4_E.get()))
    print("Passing following information to MySQL textile
    database:" +statement)
    mycursor.execute(statement)
    pycon.commit() pycon.close()
    print('Done!!')

#创建窗口实例
mw = Tk()
mw.geometry('400x400')
table_name = Label(mw, text = 'TABLE NAME')
table_name.place(x =10, y =10)
table_name_E = Entry(mw,bd = 2)
table_name_E.place(x = 150, y =10)
```

```
column1 = Label(mw, text = 'FABRIC_NAME')
column1.place(x = 10 , y = 60)

column1_E = Entry(mw,bd = 2)
column1_E.place(x = 150 , y = 60 )

column2 = Label(mw, text = 'IMPORT_EXPORT')
column2.place(x = 10 , y = 110)

column2_E = Entry(mw,bd = 2)
column2_E.place(x = 150 , y = 110 )

column3 = Label(mw, text = 'COST_PER_METER')
column3.place(x = 10, y = 160 )

column3_E = Entry(mw,bd = 2)
column3_E.place(x = 150, y = 160 )

column4 = Label(mw, text = 'COLORS_AVAILABLE')
column4.place(x = 10, y = 210)

column4_E = Entry(mw,bd = 2)
column4_E.place(x = 150, y = 210 )

insert_button = Button(mw,text = 'INSERT VALUES', command = insert_val)
insert_button.place(x = 40 , y = 260)

mw.mainloop()
```

输出:

```
Passing the following information to MySQL textile database:
INSERT INTO FABRIC(FABRIC_NAME, IMPORT_EXPORT, COST_PER_METER, COLORS_AVAILABLE) VAlUES('COTTON','EXPORT',80,12);
Done!!
```

输出结果如图 9.5 所示。

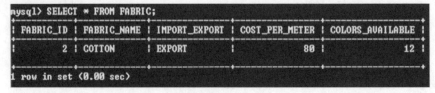

图 9.5

9.4 创建 GUI 以检索结果

创建的 GUI 如图 9.6 所示。

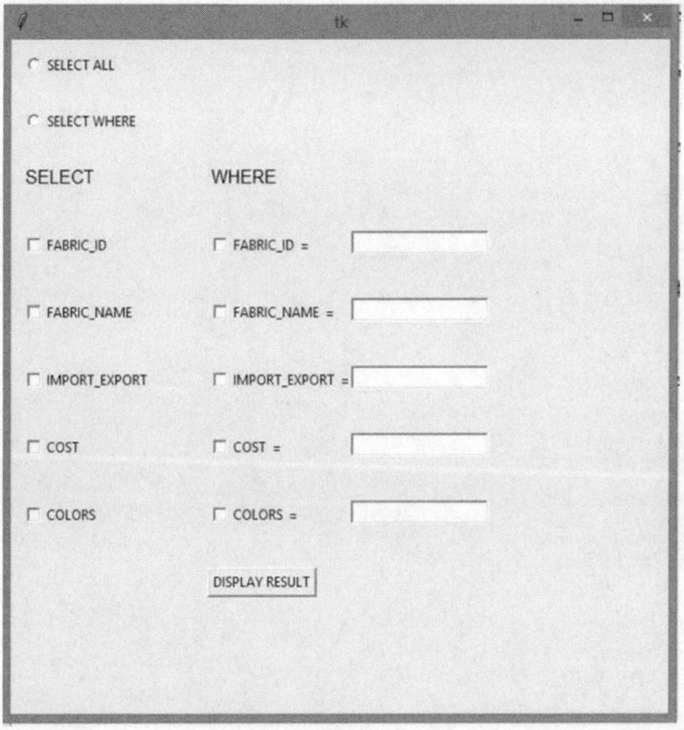

图 9.6

如果单击"SELECT ALL"单选按钮,则将显示所有记录,如图 9.7 所示。也可以按照筛选条件进行选择。

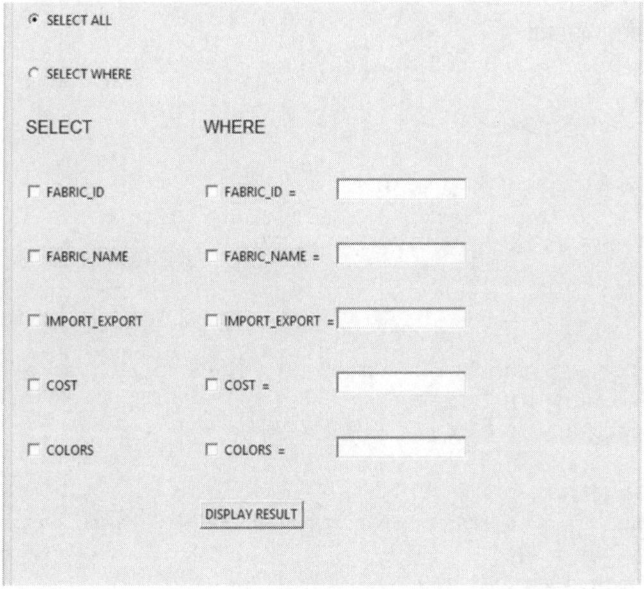

图 9.7

步骤1：编写 GUI 代码，如下所示。

```python
#导入tkinter
from tkinter import *
from tkinter import messagebox

#创建窗口实例
mw = Tk()
mw.geometry('600x600')

#单选按钮变量
var = IntVar()

#复选框变量
cvar1 = IntVar()
cvar2 = IntVar()
cvar3 = IntVar()
cvar4 = IntVar()
cvar5 = IntVar()

svar1 = IntVar()
svar2 = IntVar()
svar3 = IntVar()
svar4 = IntVar()
svar5 = IntVar()

#单选按钮 SELECT ALL
selectRadio = Radiobutton(mw, text = 'SELECT ALL',variable = var, value =1)
selectRadio.place(x =10, y =10)

#单选按钮 SELECT WHERE
selectFilterRadio = Radiobutton(mw, text = 'SELECT WHERE',variable = var, value =2)
selectFilterRadio.place(x = 10, y =60)

#标签 SELECT
selectLabel = Label(mw, text = 'SELECT',font ='Bold')
selectLabel.place(x = 10 , y = 110)

#选项 SELECT
#FABRIC_ID
fabricIdCheck = Checkbutton(mw,text ='FABRIC_ID',variable = svar1, onvalue = 1,offvalue = 0)
fabricIdCheck.place(x = 10 , y = 170 )

#FABRIC_NAME
fabricNameCheck = Checkbutton(mw,text ='FABRIC_NAME',variable = svar2, onvalue = 1,offvalue = 0)
fabricNameCheck.place(x = 10 , y = 230 )
```

```python
#IMPORT_EXPORT
ieCheck = Checkbutton(mw,text ='IMPORT_EXPORT',variable = svar3, onvalue = 1,offvalue = 0)
ieCheck.place(x = 10 , y = 290 )

#COST
costCheck = Checkbutton(mw,text ='COST ',variable = svar4, onvalue = 1, offvalue = 0)
costCheck.place(x = 10 , y = 350 )

#COLORS
colorsCheck = Checkbutton(mw, text ='COLORS',variable = svar5, onvalue = 1,offvalue = 0)
colorsCheck.place(x = 10, y = 410 )

#标签 WHERE
whereLabel = Label(mw, text = 'WHERE', font ='Bold')
whereLabel.place(x = 180 , y = 110)

#选项 WHERE
#FABRIC_ID
fabricIdCheck2 = Checkbutton(mw,text ='FABRIC_ID = ',variable = cvar1, onvalue = 1,offvalue = 0)
fabricIdCheck2.place(x = 180 , y = 170 )
fabricIdCheck2entry = Entry(mw, bd = 2)
fabricIdCheck2entry.place(x = 310, y = 170)

#FABRIC_NAME
fabricNameCheck2 = Checkbutton(mw,text ='FABRIC_NAME = ',variable = cvar2, onvalue = 1,offvalue = 0)
fabricNameCheck2.place(x = 180 , y = 230 )
fabricNameCheck2entry = Entry(mw, bd = 2)
fabricNameCheck2entry.place(x = 310, y = 230)

#IMPORT_EXPORT
ieCheck2 = Checkbutton(mw,text ='IMPORT_EXPORT = ',variable = cvar3, onvalue = 1,offvalue = 0)
ieCheck2.place(x = 180 , y = 290 )
ieCheck2entry = Entry(mw, bd = 2)
ieCheck2entry.place(x = 310, y = 290)

#COST
costCheck2 = Checkbutton(mw,text ='COST = ',variable = cvar4, onvalue = 1, offvalue = 0)
costCheck2.place(x = 180 , y = 350 )
costCheck2entry = Entry(mw, bd = 2)
costCheck2entry.place(x = 310, y = 350 )
```

```
#COLORS
colorsCheck2 = Checkbutton(mw, text = 'COLORS = ',variable = cvar5,
onvalue = 1,offvalue = 0)
colorsCheck2.place(x = 180, y = 410 )
colorsCheck2entry = Entry(mw, bd = 2)
colorsCheck2entry.place(x = 310, y = 410)
#按钮DISPLAY RESULT
display_button = Button(mw,text = 'DISPLAY RESULT')
display_button.place(x = 180 , y = 470)

mw.mainloop()
```

步骤2：编写数据库连接的导入语句，如下所示。

```
#导入mysql.connector
import mysql.connector as msql
#导入credentials.py,其中包含登录详细信息
import credentials as c
```

步骤3：设置显示数据步骤。
创建大纲。代码如下所示。

```
def display_result():
    print("var = ",var.get())
    pycon = msql.connect( ** c.dbConfig)
    mycursor = pycon.cursor()
    statement = ""

    if (var.get() == 1):
#编写代码
    elif (var.get() == 2):
#编写代码

    mycursor.execute(statement)
    result_set = mycursor.fetchall()
    for result in result_set:
        print(result)
    messagebox.showinfo("FABRIC from TEXTILE",result_set)
    pycon.close()
print('Done!! ')
```

首先为"SELECT ALL"单选按钮编写代码。如果单击此单选按钮并单击"DISPLAY RESULT"按钮，将显示表格的所有记录。在"DISPLAY RESULT"按钮的定义中，输入命令display_result并检查所有结果是否正确显示。

```
def display_result():
    print("var =",var.get())
    pycon = msql.connect(**c.dbConfig)
    mycursor = pycon.cursor()
    statement = ""

    if (var.get() == 1):
        statement = "SELECT * FROM FABRIC;"

    elif (var.get() == 2):
#编写代码
    mycursor.execute(statement)
    result_set = mycursor.fetchall()
    for result in result_set:
        print(result)
    messagebox.showinfo("FABRIC from TEXTILE",result_set)
    pycon.close()
print('Done!! ')
```

按钮代码更改如下：

```
#按钮 DISPLAY RESULT
display_button = Button(mw,text = ' DISPLAY RESULT ',command = display_result)
display_button.place(x = 180 , y = 470)
```

输出结果如图 9.8 所示。

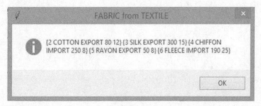

图 9.8

通过显示所有记录来检查数据库，如图 9.9 所示。

图 9.9

显示全部 5 条记录。

现在有难度的部分是根据过滤条件查询并显示某些列。

单击"SELECT WHERE"单选按钮时,根据过滤条件检查要查看的值。因此,前面研究构建查询 SELECT 部分,下面研究构建查询 WHERE 部分。代码如下。

```python
def display_result():
    print("var = ",var.get())
    pycon = msql.connect( **c.dbConfig)
    mycursor = pycon.cursor()
    statement = ""

    if (var.get() == 1):
        statement = "SELECT * FROM FABRIC;"
    elif (var.get() == 2):

        statement += "SELECT "

        if(svar1.get() == 1):
            statement += "FABRIC_ID "
        if(svar1.get() == 1 and svar2.get() == 1):
            statement += ", FABRIC_NAME "
        elif(svar1.get() == 0 and svar2.get() == 1):
            statement += "FABRIC_NAME "

        if((svar1.get() == 1 or svar2.get() == 1) and svar3.get() == 1):
            statement += ", IMPORT_EXPORT "
        elif((svar1.get() == 0 and svar2.get() == 0) and svar3.get()
        == 1):
            statement += "IMPORT_EXPORT "

        if((svar1.get() == 1 or svar2.get() == 1 or svar3.get() == 1)
        and svar4.get() == 1):
            statement + = ", COST_PER_METER "

        elif((svar1.get() == 0 and svar2.get() == 0 and svar3.get() ==
        0) and svar4.get() == 1):
            statement + = "COST_PER_METER "

        if((svar1.get() == 1 or svar2.get() == 1 or svar3.get() == 1
        or svar4.get() == 1)and svar5.get() == 1):
            statement + = " , COLORS_AVAILABLE "
        elif((svar1.get() == 0 and svar2.get() == 0 and svar3.get() ==
        0 and svar4.get() == 0)and svar5.get() == 1):
            statement + = " COLORS_AVAILABLE "

        statement + = "FROM FABRIC WHERE "
        if(cvar1.get() == 1):
            statement + = "FABRIC_ID ='{}'".format(str
            (fabricIdCheck2entry.get()))
```

```python
            if(cvar1.get() == 1 and cvar2.get() == 1):
                statement + = " AND FABRIC_NAME = '{}'".format(str
                (fabricNameCheck2entry.get()))

            elif(cvar1.get() == 0 and cvar2.get() == 1):
                statement + = " FABRIC_NAME = '{}'".format(str
                (fabricNameCheck2entry.get()))

            if((cvar1.get() == 0 and cvar2.get() == 0) and cvar3.get() == 1):
                statement + = " IMPORT_EXPORT = '{}'".format(str
                (ieCheck2entry.get()))

            elif((cvar1.get() == 1 or cvar2.get() == 1) and cvar3.get() == 1):
                statement + = " AND IMPORT_EXPORT = '{}'".format(str
                (ieCheck2entry.get()))

            if((cvar1.get() == 0 and cvar2.get() == 0 and cvar3.get() ==
            0) and cvar4.get() == 1):
                statement + = " COST_PER_METER = {} ".format(int
                (costCheck2entry.get()))

            elif((cvar1.get() == 1 or cvar2.get() == 1 or cvar3.get() ==
            1) and cvar4.get() == 1):
                statement + = " AND COST_PER_METER = {} ".format(int
                (costCheck2entry.get()))

            if((cvar1.get() == 0 and cvar2.get() == 0 and cvar3.get() == 0
            and cvar4.get() == 0)and cvar5.get() == 1):
                statement + = " COST_PER_METER = {} ".format(int
                (colorsCheck2entry.get()))
            elif((cvar1.get() == 1 or cvar2.get() == 1 or cvar3.get() == 1
            or cvar4.get() == 1)and cvar5.get() == 1):
                statement + = " AND COST_PER_METER = {} ".format(int
                (colorsCheck2entry.get()))
    statement + = ";"
    mycursor.execute(statement)
    result_set = mycursor.fetchall()
    for result in result_set:
        print(result)
    messagebox.showinfo("FABRIC from TEXTILE",result_set)
    pycon.close()
print('Done!! ')
```

SELECT 语句的语法：SELECT col1 col2…, WHERE col3 ='value' AND col6 ='value 2'。

检查界面中"SELECT"部分下被勾选的复选框，并创建查询语句，然后根据检索结果读取这些值。

最终代码如下所示。

```python
#导入tkinter
from tkinter import *
from tkinter import messagebox
#导入mysql.connector
import mysql.connector as msql
#导入credentials.py,其中包含登录详细信息
import credentials as c

def display_result():
    print("var = ",var.get())
    pycon = msql.connect(**c.dbConfig)
    mycursor = pycon.cursor()
    statement = ""

    if (var.get() == 1):
        statement = "SELECT * FROM FABRIC;"
    elif (var.get() == 2):

        statement + = "SELECT "

        if(svar1.get() == 1):
            statement + = "FABRIC_ID "

        if(svar1.get() == 1 and svar2.get() == 1):
            statement + = ", FABRIC_NAME "
        elif(svar1.get() == 0 and svar2.get() == 1):
            statement + = "FABRIC_NAME "

        if((svar1.get() == 1 or svar2.get() == 1) and svar3.get() == 1):
            statement + = ", IMPORT_EXPORT "
        elif((svar1.get() == 0 and svar2.get() == 0) and svar3.get()
            == 1):
            statement + = "IMPORT_EXPORT "

        if((svar1.get() == 1 or svar2.get() == 1 or svar3.get() == 1)
        and svar4.get() == 1):
            statement + = ", COST_PER_METER "
        elif((svar1.get() == 0 and svar2.get() == 0 and svar3.get() ==
        0) and svar4.get() == 1):
            statement + = "COST_PER_METER "

        if((svar1.get() == 1 or svar2.get() == 1 or svar3.get() == 1
        or svar4.get() == 1)and svar5.get() == 1):
            statement + = " , COLORS_AVAILABLE "
        elif((svar1.get() == 0 and svar2.get() == 0 and svar3.get() ==
        0 and svar4.get() == 0)and svar5.get() == 1):
            statement + = " COLORS_AVAILABLE "
        statement + = "FROM FABRIC WHERE "

        if(cvar1.get() == 1):
            statement + = "FABRIC_ID = '{}' ".format(str
            (fabricIdCheck2entry.get()))
```

```python
            if(cvar1.get() == 1 and cvar2.get() == 1):
                statement + = " AND FABRIC_NAME = '{}'".format(str
                    (fabricNameCheck2entry.get()))

            elif(cvar1.get() == 0 and cvar2.get() == 1):
                statement + = " FABRIC_NAME = '{}'".format(str
                    (fabricNameCheck2entry.get()))

            if((cvar1.get() == 0 and cvar2.get() == 0) and cvar3.get() == 1):
                statement + = " IMPORT_EXPORT = '{}'".format(str
                    (ieCheck2entry.get()))

            elif((cvar1.get() == 1 or cvar2.get() == 1) and cvar3.get() == 1):
                statement + = " AND IMPORT_EXPORT = '{}'".format(str
                    (ieCheck2entry.get()))

            if((cvar1.get() == 0 and cvar2.get() == 0 and cvar3.get() ==
            0) and cvar4.get() == 1):
                statement + = " COST_PER_METER = {} ".format(int
                    (costCheck2entry.get()))

            elif((cvar1.get() == 1 or cvar2.get() == 1 or cvar3.get() ==
            1) and cvar4.get() == 1):
                statement + = " AND COST_PER_METER = {} ".format(int
                    (costCheck2entry.get()))

            if((cvar1.get() == 0 and cvar2.get() == 0 and cvar3.get() == 0
            and cvar4.get() == 0)and cvar5.get() == 1):
                statement + = " COST_PER_METER = {} ".format(int
                    (colorsCheck2entry.get()))
            elif((cvar1.get() == 1 or cvar2.get() == 1 or cvar3.get() == 1
            or cvar4.get() == 1)and cvar5.get() == 1):
                statement + = " AND COST_PER_METER = {} ".format(int
                    (colorsCheck2entry.get()))

        statement + = ";"
        mycursor.execute(statement)
        result_set = mycursor.fetchall()
        for result in result_set:
            print(result)
        messagebox.showinfo("FABRIC from TEXTILE",result_set)
        pycon.close()
print('Done!! ')

#创建窗口实例
mw = Tk()
mw.geometry('600x600')
```

```python
#单选按钮变量
var = IntVar()
#复选框变量 – WHERE 部分
cvar1 = IntVar()
cvar2 = IntVar()
cvar3 = IntVar()
cvar4 = IntVar()
cvar5 = IntVar()

#复选框变量 – SELECT 部分
svar1 = IntVar()
svar2 = IntVar()
svar3 = IntVar()
svar4 = IntVar()
svar5 = IntVar()

#单选按钮 SELECT ALL
selectRadio = Radiobutton(mw, text = 'SELECT ALL',variable = var, value =1)
selectRadio.place(x =10, y =10)

#单选按钮 SELECT WHERE
selectFilterRadio = Radiobutton(mw, text = 'SELECT WHERE',variable = var, value =2)
selectFilterRadio.place(x = 10, y =60)

#标签 SELECT
selectLabel = Label(mw, text = 'SELECT',font ='Bold')
selectLabel.place(x = 10 , y = 110)

#选项 SELECT

#FABRIC_ID
fabricIdCheck = Checkbutton(mw,text ='FABRIC_ID',variable = svar1, onvalue = 1,offvalue = 0)
fabricIdCheck.place(x = 10 , y = 170 )

#FABRIC_NAME
fabricNameCheck = Checkbutton(mw,text ='FABRIC_NAME',variable = svar2, onvalue = 1,offvalue = 0)
fabricNameCheck.place(x = 10 , y = 230 )

#IMPORT_EXPORT
ieCheck = Checkbutton(mw,text ='IMPORT_EXPORT',variable = svar3, onvalue = 1,offvalue = 0)
ieCheck.place(x = 10 , y = 290 )
```

```python
#COST
    costCheck = Checkbutton(mw,text ='COST',variable = svar4,onvalue = 1,
offvalue = 0)
    costCheck.place(x = 10 , y = 350 )

    #COLORS
    colorsCheck = Checkbutton(mw, text ='COLORS',variable = svar5, onvalue
= 1,offvalue = 0)
    colorsCheck.place(x = 10, y = 410 )

    #标签 WHERE
    whereLabel = Label(mw, text = 'WHERE', font ='Bold')
    whereLabel.place(x = 180 , y = 110)

    #选项 WHERE
    #FABRIC_ID
    fabricIdCheck2 = Checkbutton(mw,text ='FABRIC_ID   = ',variable = cvar1,
onvalue = 1,offvalue = 0)
    fabricIdCheck2.place(x = 180 , y = 170 )
    fabricIdCheck2entry = Entry(mw, bd = 2)
    fabricIdCheck2entry.place(x = 310, y = 170)

    #FABRIC_NAME
    fabricNameCheck2 = Checkbutton(mw,text ='FABRIC_NAME   = ',variable =
cvar2, onvalue = 1,offvalue = 0)
    fabricNameCheck2.place(x = 180 , y = 230 )
    fabricNameCheck2entry = Entry(mw, bd = 2)
    fabricNameCheck2entry.place(x = 310, y = 230)

    #IMPORT_EXPORT
    ieCheck2 = Checkbutton(mw,text ='IMPORT_EXPORT   =  ',variable = cvar3,
onvalue = 1,offvalue = 0)
    ieCheck2.place(x = 180 , y = 290 )
    ieCheck2entry = Entry(mw, bd = 2)
    ieCheck2entry.place(x = 310, y = 290)

    #COST
    costCheck2 = Checkbutton(mw,text ='COST = ',variable = cvar4, onvalue = 1,
offvalue = 0)
    costCheck2.place(x = 180 , y = 350 )
    costCheck2entry = Entry(mw, bd = 2)
    costCheck2entry.place(x = 310, y = 350)

    #COLORS
    colorsCheck2 = Checkbutton(mw, text = 'COLORS   = ',variable = cvar5,
onvalue = 1,offvalue = 0)
```

```
colorsCheck2.place(x = 180, y = 410 )
colorsCheck2entry = Entry(mw, bd = 2)
colorsCheck2entry.place(x = 310, y = 410)

#按钮 DISPLAY RESULT
display_button = Button(mw,text = 'DISPLAY RESULT',command = display_result)
display_button.place(x = 180 , y = 470)

mw.mainloop()
```

输出结果如图9.10和图9.11所示。

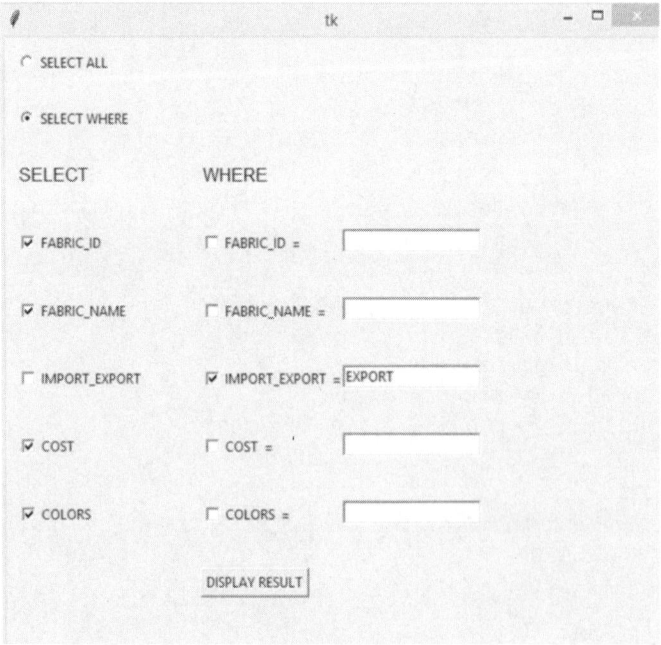

图9.10

图9.11

输出结果正确，因为IMPORT_EXPORT ='EXPORT'只有3条记录，如图9.12所示。

```
+-----------+-------------+---------------+----------------+------------------+
| FABRIC_ID | FABRIC_NAME | IMPORT_EXPORT | COST_PER_METER | COLORS_AVAILABLE |
+-----------+-------------+---------------+----------------+------------------+
|         2 | COTTON      | EXPORT        |             80 |               12 |
|         3 | SILK        | EXPORT        |            300 |               15 |
|         4 | CHIFFON     | IMPORT        |            250 |                8 |
|         5 | RAYON       | EXPORT        |             50 |                8 |
|         6 | FLEECE      | IMPORT        |            190 |               25 |
+-----------+-------------+---------------+----------------+------------------+
5 rows in set (0.00 sec)
```

图 9.12

小结

读者在第 4 章中已经学习了如何使用读者 Python 程序访问数据库，在第 8 章中创建了 GUI 表单和控件，并学习了如何使用 tkinter 模块创建 GUI。在本章中读者学习了如何使用 GUI 连接数据库。在此基础上读者已具备创建用户界面和访问数据库应用程序的全部知识，利用这些知识可以开发实时项目。

第 10 章

栈、队列和双端队列

第 10 章 栈、队列和双端队列

> **引言**　本章介绍如何使用 Python 实现栈、队列和双端队列。

> **知识结构**
> - 栈
> - 队列
>   - 基本队列函数
>   - 实现队列
>   - 使用单队列实现栈
>   - 栈的两个栈实现队列
> - 双端队列

> **目标**　完成本章的学习后，读者应了解栈、队列和双端队列及其实现。

10.1 栈

栈是有序项集合，添加项和删除项都在同一端，也称为顶部。栈的另一端称为底部。堆底很重要，因为靠近底部的项在栈中停留时间最长。最新添加的项位于顶部，因此先被删除。使用 push 操作将项添加到栈，使用 pop 操作将其删除。

为了理解 push 和 pop 操作，假设处理 Word 文档，输入标题后进行字体大小和颜色的更改，如图 10.1 所示。

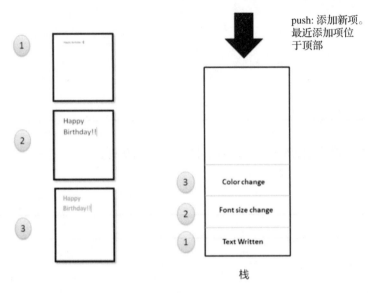

图 10.1

因此，虽然可以对 Word 文档执行任何操作，但单击"undo"按钮时，只有最后完成的任务被撤销，如图 10.2 所示。

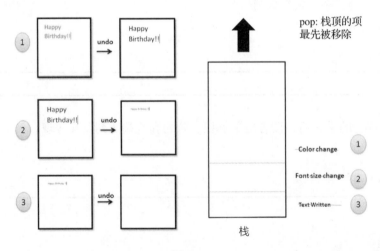

图 10.2

类似地，在栈中可以使用 push 操作添加任意项，但只弹出或删除顶部项，这意味着最后一项是第一个要删除的对象。这种排序原则称为后进先出（Last – In – First – Out，LIFO），即较新项接近顶部，而较旧项接近底部，如图 10.3 所示。

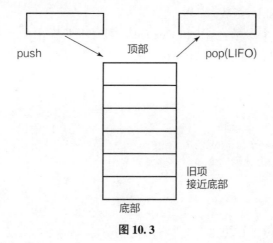

图 10.3

栈很重要，每当需要颠倒项的顺序时它都是必需的，因为移除顺序与插入顺序相反。示例如下。

（1）上网时按浏览器中的"后退"按钮。
（2）Microsoft 应用程序中的"Ctrl + Z"组合键（撤销）。
（3）缓存中删除最近的使用对象。

栈通常由软件程序员使用，使用程序时表现不太明显，因为这些操作通常在后台进行。但是，很多时候会遇到栈溢出错误。当栈的实际内存耗尽时，就会发生这种情况。

栈非常简单，它既是最重要的数据结构之一，也是数据结构和算法中非常重要的元素。

下面介绍栈的 Python 实现。

使用列表实现栈。

步骤1：定义 Stack 类。代码如下所示。

```
#定义Stack类
class Stack:
```

步骤2：创建构造函数。

创建带有参数 self 和 n（栈的大小）的构造函数。在构造函数方法中声明 self.stack 是空列表（[ ]），self.size 的值为 n。代码如下所示。

```
#定义Stack类
class Stack:

    #构造函数
    def __init__(self,n):
        self.stack = []
        self.size = n
```

步骤3：定义 push( )函数。

push( )函数有两个参数：self 和 element（列表中要弹出的元素）。在 push( )函数中，首先检查栈的长度是否等于 input(n)的大小。如果是，则表示栈满，并打印消息"栈已满，无法再追加元素"；否则调用 append( )方法将元素压入栈。代码如下所示。

```python
#push 函数
    def push(self,element):
        if len(self.stack) == self.size:
            print("栈已满,无法再追加元素")
        else:
            self.stack.append(element)
```

步骤4：定义 pop( )函数。

检查栈，如果为空，则打印"栈为空。无元素可 pop!!"；否则，从栈中弹出最后一项。代码如下所示。

```python
#pop 函数
    def pop(self):
        if(self.stack == []):
            print("栈为空。无元素可pop!!")
        else:
            self.stack.pop()
```

步骤5：编写执行代码，如下所示。

```python
s = Stack(3)
s.push(6)
s.push(2)
print(s.stack)
s.pop()
print(s.stack)
```

代码：

```python
#定义 Stack 类
class Stack:

    #构造函数
    def __init__(self, n):
        self.stack = []
        self.size = n
```

```
#push()函数
    def push(self,element):
        if len(self.stack) == self.size:
            print("no more elements can be appended as the
            stack is full")
        else:
            self.stack.append(element)
#pop()函数
    def pop(self):
        if(self.stack == []):
            print("Stack is empty. Nothing to POP!!")
        else:
            self.stack.pop()
s = Stack(3)
s.push(6)
s.push(2)
print(s.stack)
s.pop()
print(s.stack)
```

输出：

[6, 2]

[6]

>>>

示例 10.1

编写程序检查给定字符串中是否有平衡括号集。平衡括号包括()、{}、[]、{[()]}、[][]等。

答：

这里，必须检查字符串中的括号对是否以正确格式存在，例如"[]{()}"之类的表达式正确。但是，如果左括号没有相应的右括号，则括号不匹配，例如"[}"或"{}[))"。按照以下步骤解决此问题。

步骤1： 定义 paranthesis_match 类。代码如下所示。

```
class paranthesis_match:
```

步骤2： 定义开括号和闭括号列表。

定义两个列表，使左括号索引与相应的右括号索引匹配。

（1）列表 opening_brackets 将所有类型的左括号作为元素——["(", "{", "["]。
（2）列表 closing_brackets 将所有类型的右括号作为元素——[")", "}", "]"]。

以下为定义列表的方法：

```
class paranthesis_match:
    opening_brackets = ["(","{","["]
    closing_brackets = [")","}","]"]
```

步骤3：定义构造函数、push( )函数和pop( )函数。

构造函数以字符串参数形式接收expression，用于参数验证。

初始化列表用于栈。此列表用到push( )和pop( )函数。代码如下所示。

```
#声明构造函数
def __init__(self, expression):
    self.expression = expression
    self.stack = []
```

由于使用栈实现目标，所以需要push( )和pop( )函数。

调用push( )函数时，将元素添加到栈中。代码如下所示。

```
#push 函数
def push(self,element):
    self.stack.append(element)
```

调用pop( )函数时，弹出栈中最后一个元素。代码如下所示。

```
#pop 函数
def pop(self):
    if self.stack == []:
        print("Unbalanced Paranthesis")
    else:
        self.stack.pop()
```

步骤4：定义分析函数。

现在编写代码分析字符串。

在此函数中，执行以下步骤。

（1）检查表达式长度。平衡括号字符串的字符数始终为偶数。如果表达式长度只能被2整除，则继续分析。因此，if…else循环构成该函数的外部结构。代码如下所示。

```
if len(self.expression)%2 == 0:
        ---- we analyse ----
else:
        print("Unbalanced Paranthesis")
```

（2）表达式的长度为偶数，因此继续分析，并在if块中编写代码。现在逐个元素遍历列表。如果遇到左括号，则把它推到栈上；否则，检查元素是否在closing_brackets列表中。如果元素在closing_brackets列表中，则从栈中弹出最后一个元素，并查看在列表 opening_brackets 和 closing_brackets 中元素的索引是否对应同一个括号，如果是，则存在匹配项，否则括号不平衡，如图 10.4 所示。

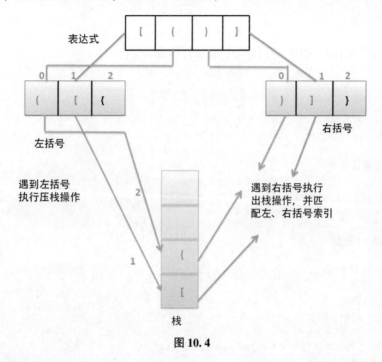

图 10.4

代码如下所示。

```
def is_match(self):

    print("expression is = ",self.expression)
    if len(self.expression)%2 == 0:
        for element in self.expression:
            print("evaluating ", element)
            if element in self.opening_brackets:
                print("it is an opening bracket - ", element,
                "pushing to stack")
                self.push(element)
                print("pushed", element, " on to stack the stack
                is ", self.stack)
            elif element in self.closing_brackets:
                x = self.stack.pop()
                print("time to pop element is ", x)
                if self.opening_brackets.index(x) == self.closing_
                brackets.index(element):
                    print("Match Found")
                else:
                    print("Match not found - check prarnthesis")
                    return;
    else:
        print("Unbalanced Paranthesis")
```

步骤5：编写执行代码，如下所示。

```
pm = paranthesis_match("([{}])")
pm.is_match()
```

因此，为便于用户使用，在代码中添加打印命令。
代码：

```python
class paranthesis_match:
    opening_brackets = ["(","{","["]
    closing_brackets = [")","}","]"]

    #声明构造函数
    def __init__(self, expression):
        self.expression = expression
        self.stack = []

    #push()函数
    def push(self,element):
        self.stack.append(element)

    #pop()函数
    def pop(self):
        if self.stack == []:
            print("Unbalanced Paranthesis")
        else:
            self.stack.pop()

    def is_match(self):

        print("expression is = ",self.expression)
        if len(self.expression)%2 == 0:
            for element in self.expression:
                print("evaluating ", element)
                if element in self.opening_brackets:
                    print("it is an opening bracket - ", element,
                     "pushing to stack")
                    self.push(element)
                    print(" pushed", element, " on to stack the stack is ",
                     self.stack)
                elif element in self.closing_brackets:
                    x = self.stack.pop()
                    print("time to pop element is ", x)
                    if self.opening_brackets.index(x) == self.closing_
                    brackets.index(element):
                        print("Match Found")
                    else:
                        print("Match not found - check prarnthesis")
                        return;
```

```
        else:
            print("Unbalanced Paranthesis")
pm = paranthesis_match("([||])")
pm.is_match()
```

输出：

```
expression is =  ([|])
evaluating  (
it is an opening bracket -  ( pushing to stack
pushed ( on to stack the stack is['(']
evaluating  [
it is an opening bracket - [ pushing to stack
pushed [  on to stack the stack is ['(', '[']
evaluating  |
it is an opening bracket - | pushing to stack
pushed |  on to stack the stack is ['(', '[', '|']
evaluating  |
time to pop element is  |
Match Found
evaluating  ]
time to pop element is  [
Match Found
evaluating  )
time to pop element is  (
Match Found
```

10.2 队列

队列的元素从一端添加，从另一端移除。队列遵循先进先出原则，从前端移除项，从后端增加项，如图 10.5 所示。因此，如同现实生活中的队列，项从后面进入队列，并随着项的逐个移除而开始向前移动。

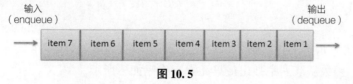

图 10.5

因此，在队列中，前面项是序列中最早添加的项，最近添加的项必须在最后等待。插入和删除操作也称为 enqueue 和 dequeue。

10.2.1 基本队列函数

基本队列函数如下。

(1) enqueue(i):将元素 i 添加到队列。
(2) dequeue():从队列中移除第一个元素并返回其值。
(3) isEmpty():布尔函数,如果队列为空,则返回 True;否则返回 False。
(4) size():返回队列的长度。

10.2.2 实现队列

编写程序实现队列。

步骤 1:定义类。代码如下所示。

```
class Queue:
```

步骤 2:定义构造函数。

初始化空列表队列。代码如下所示。

```
def __init__(self):
    self.queue =[]
```

步骤 3:定义 isEmpty() 函数。

使用 isEmpty() 函数检查队列,如果队列为空,则打印一条消息 "Queue is Empty";否则打印 "Queue is not Empty"。代码如下所示。

```
def isEmpty(self):
    if self.queue ==[]:
        print("Queue is Empty")
    else:
        print("Queue is not Empty")
```

步骤 4:定义 enqueue() 函数。

enqueue() 函数将元素作为参数并将其插入索引 0 处。所有元素在队列中移动 1 个位置。代码如下所示。

```
def enqueue(self,element):

    self.queue.insert(0,element)
```

步骤 5:定义 dequeue() 函数。

dequeue() 函数从队列中弹出最早元素。代码如下所示。

```
def dequeue(self):
    self.queue.pop()
```

步骤 6:定义 size() 函数。

size() 函数用于返回队列的长度。代码如下所示。

```python
def size(self):
    print("size of queue is",len(self.queue))
```

步骤7：编写执行代码，如下所示。

```python
#执行代码
q = Queue()
q.isEmpty()
# 插入元素
print("inserting element no.1")
q.enqueue("apple")
print("inserting element no.2")
q.enqueue("banana")
print("inserting element no.3")
q.enqueue("orange")
print("The queue elements are as follows:")
print(q.queue)
print("check if queue is empty?")
q.isEmpty()
# 移除元素
print("remove first element")
q.dequeue()
print("what is the size of the queue?")
q.size()
print("print contents of the queue")
print(q.queue)
```

代码：

```python
class Queue:
    def __init__(self):
        self.queue =[]

    def isEmpty(self):
        if self.queue ==[]:
            print("Queue is Empty")

        else:
            print("Queue is not empty")

    def enqueue(self,element):
        self.queue.insert(0,element)

    def dequeue(self):
        self.queue.pop()

    def size(self):
        print("size of queue is",len(self.queue))
```

```
#执行代码
q = Queue()
q.isEmpty()
#插入元素
print("inserting element no.1")
q.enqueue("apple")
print("inserting element no.2")
q.enqueue("banana")
print("inserting element no.3")
q.enqueue("orange")
print("The queue elements are as follows:")
print(q.queue)
print("check if queue is empty?")
q.isEmpty()
#移除元素
print("remove first element")
q.dequeue()
print("what is the size of the queue?")
q.size()
print("print contents of the queue")
print(q.queue)
```

输出:

```
Queue is Empty
inserting element no.1
inserting element no.2
inserting element no.3
The queue elements are as follows:
['orange','banana','apple']
check if queue is empty?
Queue is not empty
remove first element
what is the size of the queue?
size of queue is 2
print contents of the queue
['orange','banana']
```

10.2.3 使用单队列实现栈

在实现代码前,了解其背后的逻辑很重要。该问题让队列像栈一样工作。队列的工作原则是先进先出,而栈的工作原则是后进先出,如图 10.6 所示。

第10章 栈、队列和双端队列 355

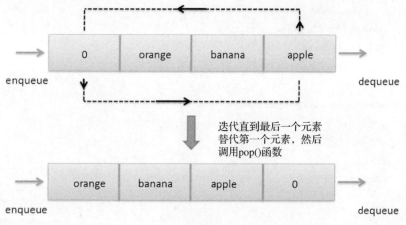

图 10.6

代码：

```
class Stack_from_Queue:
    def __init__(self):
        self.queue =[]

    def isEmpty(self):
        if self.queue ==[]:
            print("Queue is Empty")
        else:
            print("Queue is not empty")
    def enqueue(self,element):
        self.queue.insert(0,element)
    def dequeue(self):
        return self.queue.pop()
    def size(self):
        print("size of queue is",len(self.queue))

    def pop(self):
        for i in range(len(self.queue) -1):
            x = self.dequeue()
            print(x)
            self.enqueue(x)
        print("element removed is",self.dequeue())
```

执行 1：

```
sq = Stack_from_Queue()
sq.isEmpty()
print("inserting element apple")
sq.enqueue("apple")
print("inserting element banana")
```

```
sq.enqueue("banana")
print("inserting element orange")
sq.enqueue("orange")
print("inserting element 0")
sq.enqueue("0")
print("The queue elements are as follows:")
print(sq.queue)
print("check if queue is empty?")
sq.isEmpty()
print("remove the last in element")
sq.pop()
sq.pop()
sq.pop()
sq.pop()
sq.isEmpty()
```

输出1：

```
Queue is Empty
inserting element apple
inserting element banana
inserting element orange
inserting element 0
The queue elements are as follows:
['0','orange','banana','apple']
check if queue is empty?
Queue is not empty
remove the last in element
apple
banana
orange
element removed is 0
apple
banana
element removed is orange
apple
element removed is banana
element removed is apple
Queue is Empty
>>>
```

执行 2：

```
sq = Stack_from_Queue()
sq.isEmpty()
print("inserting element apple")
sq.enqueue("apple")
print("inserting element banana")
sq.enqueue("banana")
print("inserting element orange")
sq.enqueue("orange")
print("inserting element 0")
sq.enqueue("0")
for i in range(len(sq.queue)):
    print("The queue elements are as follows:")
    print(sq.queue)
    sq.pop()
    print("check if queue is empty?")
    sq.isEmpty()
    print("remove the last in element")
    print(sq.queue)
```

输出 2：

```
inserting element apple
inserting element banana
inserting element orange
inserting element 0
The queue elements are as follows:
['0','orange','banana','apple']
apple
banana
orange
element removed is 0
check if queue is empty?
Queue is not empty
remove the last in element
['orange','banana','apple']
The queue elements are as follows:
['orange','banana','apple']
apple
banana
element removed is orange
check if queue is empty?
Queue is not empty
remove the last in element
['banana','apple']
```

```
The queue elements are as follows:
['banana','apple']
apple
element removed is banana
check if queue is empty?
Queue is not empty
remove the last in element
['apple']
The queue elements are as follows:
['apple']
element removed is apple
check if queue is empty?
Queue is Empty
remove the last in element
[]
```

10.2.4 使用两个栈实现队列

下面学习如何使用两个栈实现一个队列。

步骤1：使用 push( )、pop( )和 isEmpty( )函数创建基本 Stack 类。代码如下所示。

```
class Stack:
    def __init__(self):
        self.stack []
    def push(self,element):
        self.stack.append(element)
    def pop(self):
        return self.stack.pop()
    def isEmpty(self):
        return self.stack == []
```

步骤2：定义 Queue 类。代码如下所示。

```
class Queue:
```

步骤3：定义构造函数__init__( )。这里要求定义两个栈，初始化两个栈对象。代码如下所示。

```
def __init__(self):
    self.inputStack = Stack()
    self.outputStack = Stack()
```

步骤4：定义入队 enqueue( )函数。

enqueue( )函数将元素推入第一个栈。代码如下所示。

```
def enqueue(self,element):
    self.inputStack.push(element)
```

步骤 5：定义 dequeue( ) 函数。

使用 dequeue( ) 函数检查 outputStack 是否为空。如果为空，则从 inputStack 中逐个弹出元素，并推入 outputStack，这样最后一个入元素就是第一个出元素。但是，如果 outputStack 不为空，则直接从中弹出元素。

假设插入 4 个值（1, 2, 3, 4），调用 enqueue( ) 函数，如图 10.7 所示。弹入栈过程如下：调用 dequeue( ) 函数时，inputStack 中的元素弹出并逐个被推送到 outputStack，直到最后一个元素从 inputStack 中弹出并返回。如果 outputStack 不为空，则意味着已具有正确顺序的元素，并按该顺序弹出，如图 10.8 所示。

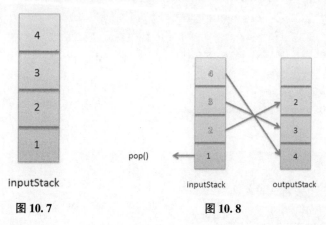

图 10.7   图 10.8

代码：

```
def dequeue(self):
    #if not self.inputStack.isEmpty():
    if self.outputStack.isEmpty():
        for i in range(len(self.inputStack.stack)-1):
            x = self.inputStack.pop()
            self.outputStack.push(x)
        print("popping out value = ", self.inputStack.pop())
    else:
        print("popping out value = ", self.outputStack.pop())
class Queue:
    def __init__(self):
        self.inputStack = Stack()
        self.outputStack = Stack()
    def enqueue(self,element):
        self.inputStack.push(element)
    def dequeue(self):
```

```
            if self.outputStack.isEmpty():
                for i in range(len(self.inputStack.stack)-1):
                    x = self.inputStack.pop()
                    self.outputStack.push(x)
                print("popping out value =",self.inputStack.pop())
            else:
                print("popping out value =",self.outputStack.pop())
#定义 Stack 类
class Stack:
    def __init__(self):
        self.stack = []
    def push(self,element):
        self.stack.append(element)
    def pop(self):
        return self.stack.pop()
        
    def isEmpty(self):
        return self.stack == []
```

执行：

```
Q = Queue()
print("insert value 1")
Q.enqueue(1)
print("insert value 2")
Q.enqueue(2)
print("insert value 3")
Q.enqueue(3)
print("insert value 4")
Q.enqueue(4)
print("dequeue operation")
Q.dequeue()
Q.dequeue()
print("insert value 7")
Q.enqueue(7)
Q.enqueue(8)
Q.dequeue()
Q.dequeue()
Q.dequeue()
Q.dequeue()
```

输出：

```
insert value 1
insert value 2
insert value 3
insert value 4
dequeue  operation
popping out value = 1
popping out value = 2
insert value 7
popping out value = 3
popping out value = 4
popping out value = 7
popping out value = 8
```

10.3 双端队列

双端队列更像队列，并且有两端，即前部和后部，如图 10.9 所示。双端队列在本质上更灵活，因为它能从前面或后面添加或删除元素。因此，这个线性数据结构具有栈和队列的优点。

图 10.9

下面学习如何编写代码实现双端队列。

实现双端队列很容易。如果从双端队列的后面添加元素，则必须在索引 0 处添加。如果从双端队列的前面添加元素，则需调用 append( ) 函数。同样，如果从前面移除元素，则需调用 pop( ) 函数；如果从后面移除元素，则需调用 pop(0) 函数。

代码：

```
class Deque:
    def __init__(self):
        self.deque =[]
    def addFront(self,element):
        self.deque.append(element)
        print("After adding from front the deque value is : ", self.deque)
    def addRear(self,element):
        self.deque.insert(0,element)
        print("After adding from end the deque value is : ", self.deque)
```

```
    def removeFront(self):
        self.deque.pop()
        print("After removing from the front the deque value is : ",
        self.deque)

    def removeRear(self):
        self.deque.pop(0)

        print("After removing from the end the deque value is : ",
        self.deque)
```

执行:

```
d = Deque()
print("Adding from front")
d.addFront(1)
print("Adding from front")
d.addFront(2)
print("Adding from Rear")
d.addRear(3)
print("Adding from Rear")
d.addRear(4)
print("Removing from Front")
d.removeFront()
print("Removing from Rear")
d.removeRear()
```

输出:

```
After adding from front the deque value is : [1]
After adding from front the deque value is : [1, 2]
After adding from end the deque value is : [3, 1, 2]
After adding from end the deque value is : [4, 3, 1, 2]
After removing from the front the deque value is : [4, 3, 1]
After removing from the end the deque value is : [3, 1]
```

第11章 链表

引言　本章介绍如何使用线性数据结构——Python 链表。

知识结构
- 链表简介
- 实现节点类
  - 遍历链表
  - 在链表头添加节点
  - 在链表尾添加节点
  - 在两个节点间插入节点
  - 从链表中删除节点
  - 打印链表的中心节点值
  - 实现双向链表
  - 反向链表

目标　完成本章的学习后，读者应全面了解链表及其用法。

11.1 链表简介

链表是由元素组成的线性结构，每个元素都是单独对象，包含数据和指针信息。

链表中的每个元素称为节点。图 11.1 中的第 1 个节点称为 Head（头部），是链表的入口点。如果链表为空，则 Head 指向 None。链表的最后一个节点指向 None。

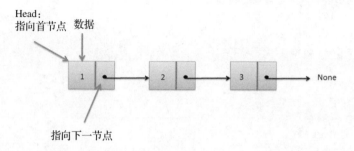

图 11.1

节点的数量可以根据需要增减，链表是动态数据结构。但是，在链表中无法直接访问数据。搜索任何项都要从 Head 开始，必须遍历每个引用才能获得该项。链表占用更多内存。

上述链表称为单链表。还有一种链表称为双链表。双链表具有对下一个节点和上一个节点的指针，如图 11.2 所示。

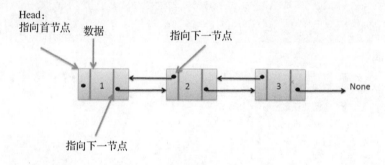

图 11.2

11.2 实现节点类

本节介绍如何实现节点类。节点包含数据和指针对下一个节点的引用。

实现过程：创建节点对象，将数据值传递给构造函数。构造函数将数值赋给数据，并将节点对象的引用设置为 None。创建所有节点对象后，将第 2 个节点对象的内存地址指向对第 1 个节点对象的引用，将第 3 个节点对象的内存地址指向对第 2 个节点对象的引用，依此类推。因此，最后一个节点对象没有引用。

实现节点类的代码如下。

代码：

```
class Node:
    def __init__(self,data = None):
        self.data = data
        self.reference = None
```

执行：

```
objNode1 = Node(1)
objNode2 = Node(2)
objNode3 = Node(3)
objNode4 = Node(4)
objNode1.reference = objNode2
objNode2.reference = objNode3
objNode3.reference = objNode4
objNode4.reference = None

print("DATA VALUE  =  ",objNode1.data,"REFERENCE  =  ",objNode1.reference)
print("DATA VALUE  =  ",objNode2.data,"REFERENCE  =  ",objNode2.reference)
print("DATA VALUE  =  ",objNode3.data,"REFERENCE  =  ",objNode3.reference)
print("DATA VALUE  =  ",objNode4.data,"REFERENCE  =  ",objNode4.reference)
```

最终代码：

```
class Node:
    def __init__(self,data = None):
        self.data = data
        self.reference = None
#执行
objNode1 = Node(1)
objNode2 = Node(2)
objNode3 = Node(3)
objNode4 = Node(4)
objNode1.reference = objNode2
objNode2.reference = objNode3
objNode3.reference = objNode4
objNode4.reference = None

print("DATA VALUE = ",objNode1.data,"REFERENCE = ",objNode1.reference)
print("DATA VALUE = ",objNode2.data,"REFERENCE = ",objNode2.reference)
print("DATA VALUE = ",objNode3.data,"REFERENCE = ",objNode3.reference)
print("DATA VALUE = ",objNode4.data,"REFERENCE = ",objNode4.reference)
```

输出：

```
DATA VALUE =   1 REFERENCE = <__main__.Node object at 0x0284B490>
DATA VALUE =   2 REFERENCE = <__main__.Node object at 0x0284B448>
DATA VALUE =   3 REFERENCE = <__main__.Node object at 0x0284B3A0>
DATA VALUE =   4 REFERENCE = None
>>>
```

11.2.1 遍历链表

下面学习如何遍历链表。

1. 方法1

为 Node 类编写代码，如下所示。

```
class Node:
    def __init__(self,data = None):
        self.data = data
        self.reference = None

objNode1 = Node(1)
objNode2 = Node(2)
objNode3 = Node(3)
objNode4 = Node(4)
```

遍历链表的步骤如下。

步骤1：创建变量 presentNode 并将第一个节点对象分配给它。代码如下所示。

```
presentNode = objNode1
```

执行此操作时，变量 presentNode 读取 objNode1 的数据和指针值。

步骤2：引用指向 objNode2。代码如下所示。

```
while presentNode:
    print("DATA VALUE = ",presentNode.data)
    presentNode = presentNode.reference
```

分配变量 presentNode 之后，objNode4 包含指针值。它将退出 while 循环，因为指针值为 None。

代码：

```
class Node:
    def __init__(self,data = None):
        self.data = data
        self.reference = None
```

执行:

```
objNode1 = Node(1)
objNode2 = Node(2)
objNode3 = Node(3)
objNode4 = Node(4)
objNode1.reference = objNode2
objNode2.reference = objNode3
objNode3.reference = objNode4
objNode4.reference = None
presentNode = objNode1
while presentNode:
    print("DATA VALUE = ",presentNode.data)
    presentNode = presentNode.reference
```

最终代码:

```
class Node:
    def __init__(self,data = None):
        self.data = data
        self.reference = None
objNode1 = Node(1)
objNode2 = Node(2)
objNode3 = Node(3)
objNode4 = Node(4)
objNode1.reference = objNode2
objNode2.reference = objNode3
objNode3.reference = objNode4
objNode4.reference = None
presentNode = objNode1
while presentNode:
    print("DATA VALUE = ",presentNode.data)
    presentNode = presentNode.reference
```

输出:

```
DATA VALUE = 1
DATA VALUE = 2
DATA VALUE = 3
DATA VALUE = 4
```

2. 方法2

创建两个类：节点和链表。

代码:

```
class Node:
    def __init__(self,data = None):
        self.data = data
        self.reference = None
```

```python
class Linked_list:
    def __init__(self):
        self.head = None
    def traverse(self):
        presentNode = self.head
        while presentNode:
            print("DATA VALUE = ",presentNode.data)
            presentNode = presentNode.reference
```

执行:

```python
objNode1 = Node(1)
objNode2 = Node(2)
objNode3 = Node(3)
objNode4 = Node(4)
linkObj = Linked_list()
#指向第一个节点对象的链表的开头
linkObj.head = objNode1
#第一个节点对象对第二个节点对象的引用
linkObj.head.reference = objNode2
objNode2.reference = objNode3
objNode3.reference = objNode4
linkObj.traverse()
```

输出:

```
DATA VALUE = 1
DATA VALUE = 2
DATA VALUE = 3
DATA VALUE = 4
```

11.2.2 在链表头添加节点

链表头添加节点，添加新节点即可。在上一节示例中添加的代码如下所示。

```python
linkObj.head = objNode1
```

如果要在链表头添加节点，只需令 linkObj. head = new_node 和 new node. reference = obj_Node1。

为此，编写代码，首先将 linkObj. head 值传递给 new node. reference，然后将 linkObj. head 设置为新节点对象。代码如下所示。

```python
def insert_at_Beginning(self,data):
    new_data = Node(data)
    new_data.reference = self.head
    self.head = new_data
```

代码：

```
class Node:
    def __init__(self,data = None):
        self.data = data
        self.reference = None
class Linked_list:
    def __init__(self):
        self.head = None
    def traverse(self):
        presentNode = self.head
        while presentNode:
            print("DATA VALUE = ",presentNode.data)
            presentNode = presentNode.reference
    def insert_at_Beginning(self,data):
        new_data = Node(data)
        new_data.reference = self.head
        self.head = new_data
```

执行：

```
objNode1 = Node(1)
objNode2 = Node(2)
objNode3 = Node(3)
objNode4 = Node(4)
linkObj = Linked_list()
#指向第一个节点对象的链表的开头
linkObj.head = objNode1
#第一个节点对象对第二个节点对象的引用
linkObj.head.reference =   objNode2
objNode2.reference = objNode3
objNode3.reference = objNode4
linkObj.insert_at_Beginning(5)
linkObj.traverse()
```

输出：

```
DATA VALUE = 5
DATA VALUE = 1
DATA VALUE = 2
DATA VALUE = 3
DATA VALUE = 4
```

11.2.3 在链表尾添加节点

在链表尾添加节点，关键是将最后一个节点的指针指向新节点。

步骤1：定义函数。代码如下所示。

```
def insert_at_end(self,data):
```

步骤 2：创建新的 Node 对象。代码如下所示。

```
new_data = Node(data)
```

步骤 3：遍历链表到最后一个节点。

不能直接访问链表的最后一个节点。必须遍历所有节点并到最后一个节点时才能进行下一步。代码如下所示。

```
presentNode = self.head
    while presentNode.reference ! = None:
        presentNode = presentNode.reference
```

步骤 4：在链表末尾添加新节点。

遍历链表后，当 presentNode.reference = None 时，就到达最后一个节点。因为不是最后一个节点，所以需要执行以下操作：

```
presentNode.reference = new_data
```

代码：

```
class Node:

    def __init__(self,data = None):
        self.data = data
        self.reference = None

class Linked_list:
    def __init__(self):
        self.head = None

    def traverse(self):
        presentNode = self.head
        while presentNode:
            print("DATA VALUE = ",presentNode.data)
            presentNode = presentNode.reference

    def insert_at_end(self,data):
        new_data = Node(data)
        presentNode = self.head
        while presentNode.reference ! = None:
            presentNode = presentNode.reference
        presentNode.reference = new_data
```

执行：

```
objNode1 = Node(1)
objNode2 = Node(2)
objNode3 = Node(3)
```

```
objNode4 = Node(4)
linkObj = Linked_list()

#指向第一个节点对象的链表的开头
linkObj.head = objNode1

#第一个节点对象对第二个节点对象的引用
linkObj.head.reference = objNode2
objNode2.reference = objNode3
objNode3.reference = objNode4

linkObj.insert_at_end(5)
linkObj.insert_at_end(6)
linkObj.insert_at_end(7)
linkObj.traverse()
```

输出:

```
DATA VALUE =   1
DATA VALUE =   2
DATA VALUE =   3
DATA VALUE =   4
DATA VALUE =   5
DATA VALUE =   6
DATA VALUE =   7
```

11.2.4 在两个节点间插入节点

在两个节点间插入节点的解决方法类似在链表头添加节点。唯一的区别是，在链表头添加节点时，要将链表头的引用值指向新节点。插入节点时，需要定义一个函数，该函数用到两个参数：节点对象，即插入的新对象；新对象数据。创建新节点后，将存储在现有节点对象中的指针值传递给新节点，然后让现有节点指向新节点对象。

步骤 1：定义函数。代码如下所示。

```
def insert_in_middle(self,insert_data,new_data):
```

步骤 2：分配引用值。代码如下所示。

```
new_node = Node(new_data)
    new_node.reference = insert_data.reference
    insert_data.reference = new_node
```

代码:

```python
class Node:
    def __init__(self,data = None):
        self.data = data
        self.reference = None

class Linked_list:
    def __init__(self):
        self.head = None

    def traverse(self):
        presentNode = self.head
        while presentNode:
            print("DATA VALUE = ",presentNode.data)
            presentNode = presentNode.reference

    def insert_in_middle(self,insert_data,new_data):
        new_node = Node(new_data)
        new_node.reference = insert_data.reference
        insert_data.reference = new_node
```

执行:

```python
objNode1 = Node(1)
objNode2 = Node(2)
objNode3 = Node(3)
objNode4 = Node(4)
linkObj = Linked_list()
#指向第一个节点对象的链表的开头
linkObj.head = objNode1
#第一个节点对象对第二个节点对象的引用
linkObj.head.reference = objNode2
objNode2.reference = objNode3
objNode3.reference = objNode4
linkObj.insert_in_middle(objNode3,8)
linkObj.traverse()
```

输出:

```
DATA VALUE =  1
DATA VALUE =  2
DATA VALUE =  3
DATA VALUE =  8
DATA VALUE =  4
>>>
```

11.2.5 从链表中删除节点

假设链表如下:

A→B→C

A. reference ＝ B

B. reference ＝ C

C. reference ＝ A.

如果要删除节点 B，则需遍历链表。当到达指针指向节点 B 的节点 A 时，将该值替换为存储在节点 B 中的指针值（指向节点 C）。这样将使节点 A 指向节点 C，从而将节点 B 从链表中删除。

remove( )函数的代码如下所示。

```python
def remove(self,removeObj):
    presentNode = self.head
    while presentNode:
        if(presentNode.reference == removeObj):
            presentNode.reference = removeObj.reference
        presentNode = presentNode.reference
```

remove( )函数以 Node 对象为参数，遍历链表，直到到达需要移除的对象。一旦指针到达必须删除的节点，只需将指针值更改为存储在对象 removeObj 中的指针值。因此，节点现在直接指向 removeObj 之后的节点。

代码：

```python
class Node:
    def __init__(self,data = None):
        self.data = data
        self.reference = None

class Linked_list:
    def __init__(self):
        self.head = None
    def traverse(self):
        presentNode = self.head
        while presentNode:
            print("DATA VALUE = ",presentNode.data)
            presentNode = presentNode.reference

    def remove(self,removeObj):
        presentNode = self.head
        while presentNode:
            if(presentNode.reference == removeObj):
                presentNode.reference = removeObj.reference
            presentNode = presentNode.reference
```

执行：

```
objNode1 = Node(1)
objNode2 = Node(2)
objNode3 = Node(3)
objNode4 = Node(4)
linkObj = Linked_list()
#指向第一个节点对象的链表的开头
linkObj.head = objNode1
#第一个节点对象对第二个节点对象的引用
linkObj.head.reference = objNode2
objNode2.reference = objNode3
objNode3.reference = objNode4
linkObj.remove(objNode2)
linkObj.traverse()
```

输出：

```
DATA VALUE =  1
DATA VALUE =  3
DATA VALUE =  4
>>>
```

11.2.6 打印链表的中心节点值

打印链表的中心节点值需要计算节点数，如果节点数是偶数，则打印中间两个节点值；否则，只打印中间节点值。

步骤1：定义函数。代码如下所示。

```
def find_middle(self,llist):
```

步骤2：求计数器长度。

设置变量 counter = 0。遍历链表时，递增计数器。在 while 循环结束时，得到链表中的节点数，即链表长度。代码如下所示。

```
counter = 0
        presentNode = self.head
        while presentNode:
            presentNode = presentNode.reference
            counter = counter + 1
        print("size of linked list = ",counter)
```

步骤3：到达链表中间。

将对中间节点的引用存储在之前的节点中。因此，在 for 循环中，迭代（counter/2）或（counter-1）/2 次。最终到达中心节点之前的节点。代码如下所示。

```
presentNode = self.head
        for i in range((counter-1)//2):
            presentNode = presentNode.reference
```

步骤4：根据链表中的节点数决定是否显示结果。

如果链表有偶数个节点，则打印存储在当前节点和下一个节点中的值。代码如下所示。

```
if (counter%2 == 0):
        nextNode = presentNode.reference
        print("Since the length of linked list is an even number the two midle elements are:")
        print(presentNode.data,nextNode.data)
```

否则，打印当前节点值。代码如下所示。

```
else:
        print("Since the length of the linked list is an odd number, the middle element is: ")
        print(presentNode.data)
```

代码：

```
class Node:
    def __init__(self,data = None):
        self.data = data
        self.reference = None
class Linked_list:
    def __init__(self):
        self.head = None

    def find_middle(self,llist):
        counter = 0
        presentNode = self.head
        while presentNode:
            presentNode = presentNode.reference
            counter = counter + 1
        print("size of linked list = ",counter)
        presentNode = self.head

        for i in range((counter-1)//2):
            presentNode = presentNode.reference
        if (counter%2 == 0):
            nextNode = presentNode.reference
            print("Since the length of linked list is an even number the two midle elements are:")
            print(presentNode.data,nextNode.data)
        else:
            print("Since the length of the linked list is an odd number, the middle element is: ")
            print(presentNode.data)
```

执行（奇数个节点）：

```
objNode1 = Node(1)
objNode2 = Node(2)
objNode3 = Node(3)
objNode4 = Node(4)
objNode5 = Node(5)
linkObj = Linked_list()
#指向第一个节点对象的链表的开头
linkObj.head = objNode1
#第一个节点对象对第二个节点对象的引用
linkObj.head.reference = objNode2
objNode2.reference = objNode3
objNode3.reference = objNode4
objNode4.reference = objNode5
linkObj.find_middle(linkObj)
```

输出：

```
size of linked list = 5

Since the length of the linked list is an odd number, the middle element is:

3
```

执行（偶数个节点）：

```
objNode1 = Node(1)
objNode2 = Node(2)
objNode3 = Node(3)
objNode4 = Node(4)
linkObj = Linked_list()
#指向第一个节点对象的链表的开头
linkObj.head = objNode1
#第一个节点对象对第二个节点对象的引用
linkObj.head.reference = objNode2
objNode2.reference = objNode3
objNode3.reference = objNode4
linkObj.find_middle(linkObj)
```

输出：

```
size of linked list = 4
Since the length of linked list is an even number the two midle elements are:
2 3
```

11.2.7　实现双向链表

双向链表包含 3 个部分：指向前一个节点的指针、数据和指向下一个节点的指针。双向链表的实现很容易，只需处理一件事，即每个节点都连接到下一个和上一个数据。

步骤 1：创建 Node 类。

使用 Node 类构造函数__init__()初始化 3 个参数：数据（data）、指向下一个节点的指针（refNext）和指向前一个节点的指针（refPrev）。代码如下所示。

```
class Node:
    def __init__(self,data = None):
        self.data = data
        self.refNext = None
        self.refPrev = None
```

步骤 2：定义函数遍历双向链表。

（1）向前遍历。

refNext 指向链表下一个值以向前遍历。从头开始，使用 refNext 移动到下一个节点。代码如下所示。

```
def traverse(self):
    presentNode = self.head
    while presentNode:
        print("DATA VALUE = ",presentNode.data)
        presentNode = presentNode.refNext
```

（2）向后遍历。

向后遍历与向前遍历相反。使用 refPrev 向后遍历，因为它指向前一个节点。从尾部开始，使用 refPrev 移动到上一个节点。代码如下所示。

```
def traverseReverse(self):
    presentNode = self.tail
    while presentNode:
        print("DATA VALUE = ",presentNode.data)
        presentNode = presentNode.refPrev
```

步骤 3：编写 append()函数，在双向链表尾添加节点。

在双向链表尾追加节点与在链表中追加节点相同，唯一的区别是确保追加节点的 refPrev 指向其之后添加的节点。代码如下所示。

```
def append(self,data):
    new_data = Node(data)
    presentNode = self.head
    while presentNode.refNext != None:
        presentNode = presentNode.refNext
    presentNode.refNext = new_data
    new_data.refPrev = presentNode
```

步骤4：编写 remove( ) 函数删除节点。

remove( ) 函数将需要移除的节点对象作为参数。为了删除节点，遍历双向链表两次。首先从链表头开始使用 refNext 向前遍历，遇到删除对象时，将当前节点（当前指向删除对象）的 refNext 值更改为删除对象之后的节点。接着从链表尾开始向后遍历，当再次遇到删除对象时，将当前节点的 refPrev 值更改为其之前的节点。代码如下所示。

```
def remove(self,removeObj):
    presentNode = self.head
    presentNodeTail = self.tail
    while presentNode.refNext != None:
        if(presentNode.refNext == removeObj):
            presentNode.refNext = removeObj.refNext
        presentNode = presentNode.refNext
    while presentNodeTail.refPrev != None:
        if(presentNodeTail.refPrev == removeObj):
            presentNodeTail.refPrev = removeObj.refPrev
        presentNodeTail = presentNodeTail.refPrev
```

代码：

```
class Node:
    def __init__(self,data = None):
        self.data = data
        self.refNext = None
        self.refPrev = None
class dLinked_list:
    def __init__(self):
        self.head = None
        self.tail = None

    def append(self,data):
        new_data = Node(data)
        presentNode = self.head
        while presentNode.refNext != None:
            presentNode = presentNode.refNext
        presentNode.refNext = new_data
        new_data.refPrev = presentNode
        self.tail = new_data
```

```python
    def traverse(self):
        presentNode = self.head
        while presentNode:
            print("DATA VALUE = ",presentNode.data)
            presentNode = presentNode.refNext

    def traverseReverse(self):
        presentNode = self.tail
        while presentNode:
            print("DATA VALUE = ",presentNode.data)
            presentNode = presentNode.refPrev

    def remove(self,removeObj):
        presentNode = self.head
        presentNodeTail = self.tail
        while presentNode.refNext! = None:
            if(presentNode.refNext == removeObj):
                presentNode.refNext = removeObj.refNext
            presentNode = presentNode.refNext
        while presentNodeTail.refPrev! = None:
            if(presentNodeTail.refPrev == removeObj):
                presentNodeTail.refPrev = removeObj.refPrev
            presentNodeTail = presentNodeTail.refPrev
```

执行：

```
objNode1 = Node(1)
objNode2 = Node(2)
objNode3 = Node(3)
objNode4 = Node(4)
dlinkObj = dLinked_list()
#指向第一个节点对象的链表的开头
dlinkObj.head = objNode1
dlinkObj.tail = objNode4
#第一个节点对象对第二个节点对象的引用
dlinkObj.head.refNext = objNode2
dlinkObj.tail.refPrev = objNode3
objNode2.refNext = objNode3
objNode3.refNext = objNode4
objNode4.refPrev = objNode3
objNode3.refPrev = objNode2
objNode2.refPrev = objNode1
print("Appending Values")
dlinkObj.append(8)
dlinkObj.append(9)
print("traversing forward after Append")
```

```
dlinkObj.traverse()
print("traversing reverse after Append")
dlinkObj.traverseReverse()
print("Removing Values")
dlinkObj.remove(objNode2)
print("traversing forward after Remove")
dlinkObj.traverse()
print("traversing reverse after Remove")
dlinkObj.traverseReverse()
```

输出：

```
Appending Values
traversing forward after Append
DATA VALUE =  1
DATA VALUE =  2
DATA VALUE =  3
DATA VALUE =  4
DATA VALUE =  8
DATA VALUE =  9
traversing reverse after Append
DATA VALUE =  9
DATA VALUE =  8
DATA VALUE =  4
DATA VALUE =  3
DATA VALUE =  2
DATA VALUE =  1
Removing Values
traversing forward after Remove
DATA VALUE =  1
DATA VALUE =  3
DATA VALUE =  4
DATA VALUE =  8
DATA VALUE =  9
traversing reverse after Remove
DATA VALUE =  9
DATA VALUE =  8
DATA VALUE =  4
DATA VALUE =  3
DATA VALUE =  1
>>>
```

11.2.8 反向链表

反向链表必须反转指针。图 11.3 中的第 1 张表显示链表信息的存储，第 2 张表显示在开始遍历列表和反转元素前 reverse( ) 函数初始化参数的过程。

node 1		node 2		node 3		node 4	
data	reference to	data	reference to	data	reference to	data	reference to
1	node2	2	node3	3	node4	4	none

初始化

Parameters	set to value of	Final Value
previous	None	None
presentNode	self.head	node1
nextval	presentNode.refNext	node2

图 11. 3

然后，使用 while 循环，如下所示。

```
while nextval != None:
    presentNode.refNext = previous
    previous = presentNode
    presentNode = nextval
    nextval = nextval.refNext
presentNode.refNext = previous
self.head = presentNode
```

图 11. 4 所示是 while 循环的工作原理。

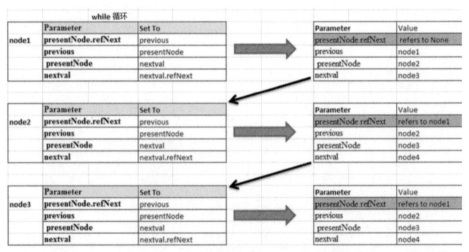

图 11. 4

遍历 while 循环时，观察 presentNode 值的变化。参考下一个变化，之前指向 node2 的 node1 将其指针更改为 None。以同样方式 node2 将其指针值更改为 node1，依此类推。

代码：

```
class Node:

    def __init__(self,data = None):
        self.data = data
        self.refNext = None
```

```python
class Linked_list:
    def __init__(self):
        self.head = None

    def reverse(self):
        previous = None
        presentNode = self.head
        nextval = presentNode.refNext
        while nextval ! = None:
            presentNode.refNext = previous
            previous = presentNode
            presentNode = nextval
            nextval = nextval.refNext

        presentNode.ref4 = previous
        self.head = presentNode

    def traverse(self):

        presentNode = self.head
        while presentNode:
            print("DATA VALUE = ",presentNode.data)
            presentNode = presentNode.refNext
```

执行：

```
objNode1 = Node(1)
objNode2 = Node(2)
objNode3 = Node(3)
objNode4 = Node(4)
linkObj = Linked_list()
#指向第一个节点对象的链表的开头
linkObj.head = objNode1
#第一个节点对象对第二个节点对象的引用
linkObj.head.refNext = objNode2
objNode2.refNext = objNode3
objNode3.refNext = objNode4
print("traverse before reversing")
linkObj.traverse()
linkObj.reverse()
print("traverse after reversing")
linkObj.traverse()
```

输出：

```
traverse before reversing
DATA VALUE =   1
DATA VALUE =   2
DATA VALUE =   3
DATA VALUE =   4
traverse after reversing
DATA VALUE =   4
DATA VALUE =   3
DATA VALUE =   2
DATA VALUE =   1
```

小结　　Python 标准库不提供链表。链表按顺序包含数据元素，这些数据元素通过链接相互连接。链表元素易于删除或插入，而不影响其基本结构。链表是动态数据结构，运行时能收缩或增长，但是链表会占用更多内存并且遍历计算量大。

12

第12章

树

> **引言**　本章介绍如何实现层次数据结构——树。

> **知识结构**
> - 引言
> - 简单树表示法
> - 树的列表表示
> - 二叉堆

> **目标**　完成本章的学习后，读者应深入掌握树的使用方法。

12.1 引言

前面章节介绍的数据结构类型用于解决应用程序问题。虽然读者已经理解了链表、栈和队列的工作方式及其在编写应用程序中的具体使用，但它们是非常受限的数据结构。处理线性数据结构是最大问题，如果必须进行搜索，则所花费时间会随着数据大小而线性增加。在某些情况下，线性结构确实有用，但事实上，对于高速情况，它们并不是好的选择。

现在从线性数据结构转移到称为树的非线性数据结构。每棵树都有一个称为根的节点。与现实生活中的树不同，树的数据结构从父节点向下分支到子节点，除根节点外，每个节点都与另一个节点直接相连。

从图 12.1 中不难看出以下事实。

A 是根节点，是 3 个节点（B、C 和 D）的父节点。同理，B 是 E 的父节点，C 是 F 和 G 的父节点。没有子节点的 D、E、F 和 G 称为叶节点或外部节点。至少有子节点（如 B 和 C）的节点称为内部节点。从根到节点的边数称为节点的深度或层次。B 的深度是 1，而 G 的深度是 2。

节点的高度是从节点到最深叶子的边数。B、C 和 D 是兄弟姐妹，因为它们有相同的父节点 A。类似地，F 和 G 也是兄弟姐妹，因为它们有相同的父节点 C。节点的子节点与另外节点的子节点相互独立。

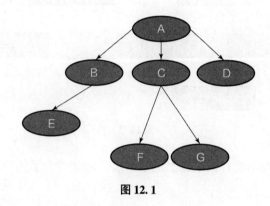

图 12.1

1. 每个叶节点唯一

（1）计算机的文件系统是树结构的典型示例。

（2）节点的附加信息称为播放负载。播放负载在算法中不重要，但在现代计算机应用中非常重要。

（3）边将两个节点连接，表示其关系。

（4）每个节点（根节点除外）只有一条入边。然而，节点可能有几条出边。

（5）根是树中唯一无入边的节点，标志树的起点。

（6）来自同一节点的入边的节点集合是该节点的子节点。

（7）节点是所有通过出边连接到其节点的父节点。

（8）父节点及其所有子节点组成的一组节点和边称为子树。

（9）根到每个节点的路径唯一。

（10）最多有两个子树的树称为二叉树。

2. 树的递归定义

树可以为空，也可以有 1 个根、0 个或多个子树。每个子树的根通过一条边连接到父树的根。

12.2　简单树表示法

观察图 12.2，这里考虑二叉树的情况。

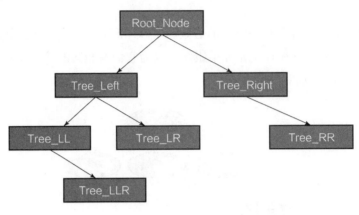

图 12.2

在二叉树中，节点不能有两个以上的子节点。因此，为了便于理解，前面的场景类似图 12.3 所示。

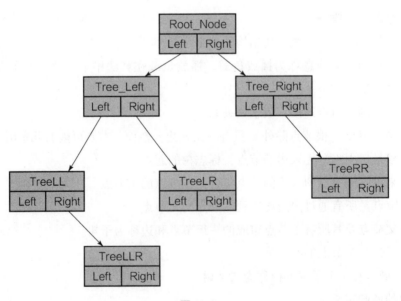

图 12.3

在图 12.3 中，左边和右边分别是引用位于该节点的左边和右边的节点实例。每个节点都有 3 个值：数据、引用左孩、引用右孩。

因此，Node 类中创建 Node 对象的构造函数如图 12.4 所示。

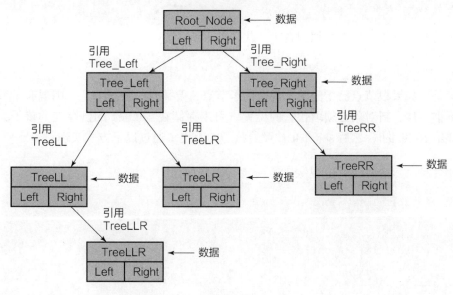

图 12.4

现在创建 Node 类。代码如下所示。

```
Class Node(object):
    def __init__(self,data_value):
        self.data_value = data_value
        self.left = None
        self.right = None
```

然后，创建根节点。代码如下所示。

```
# Root_Node
print("Create Root Node")
root = Node("Root_Node")
print("Value of Root = ",root.data_value," left = ",root.left," right = ",root.right)
```

执行这段代码，输出如下所示。

```
Value of Root = Root_Node left = None right = None
Value of Node = Tree_Left left = None right = None
```

现在，编写代码将值插入左边或右边。

当节点被创建时，最初它的左、右引用指向 None。

要在左边添加子元素。代码如下所示。

```
self.left = child_node
```

子元素可以用类似方式加到右边。代码如下所示。

```
self.right = child_node
```

但是，如果根节点已指向某个现有的子节点，要插入一个子节点，则现有子节点应该被下推一层，新对象必须取代它的位置。引用存储在 self.left 的已存在的孩子，并将其传递给 child.left，然后 self.left 被赋给孩子的引用。通过以下方式实现：

```
def insert_left(self, child):
    if self.left is None:
        self.left = child
    else:
        child.left = self.left
        self.left = child

def insert_right(self, child):
    if self.right is None:
        self.right = child
    else:
        child.right = self.right
        self.right = child
```

代码：

```
class Node(object):
    def __init__(self, data_value):
        self.data_value = data_value
        self.left = None
        self.right = None

    def insert_left(self, child):
        if self.left is None:
            self.left = child
        else:
            child.left = self.left
            self.left = child

    def insert_right(self, child):
        if self.right is None:
            self.right = child
        else:
            child.right = self.right
            self.right = child
```

执行：

```python
# Root_Node
print("Create Root Node")
root = Node("Root_Node")
print("Value of Root = ",root.data_value," left =",root.left, " right = ",root.right)

#Tree_Left
print("Create Tree_Left")
tree_left = Node("Tree_Left")
root.insert_left(tree_left)
print("Value of Node = ",tree_left.data_value," left =",tree_left.left," right = ",tree_left.right)
print("Value of Root = ",root.data_value," left =",root.left, " right = ",root.right)

#Tree_Right
print("Create Tree_Right")
tree_right = Node("Tree_Right")
root.insert_right(tree_right)
print("Value of Node = ",tree_right.data_value," left =",tree_right.left," right = ",tree_right.right)
print("Value of Root = ",root.data_value," left =",root.left, " right = ",root.right)

#TreeLL
print("Create TreeLL")
treell = Node("TreeLL")
tree_left.insert_left(treell)
print("Value of Node = ",treell.data_value," left =",treell.left,"right = ",treell.right)
print("Value of Node = ",tree_left.data_value," left =",tree_left.left," right = ",tree_left.right)
print("Value of Root = ",root.data_value," left =",root.left," right = ",root.right)
```

输出：

```
Create Root Node
Value of Root = Root_Node  left = None  right = None
Create Tree_Left
Value of Node = Tree_Left  left = None  right = None
Value of Root = Root_Node  left = <__main__.Node object at 0x000000479EC84F60>  right = None
Create Tree_Right
```

```
Value of Node =  Tree_Right  left = None  right = None
Value  of  Root  =  Root_Node   left  =  <__main__.Node object  at
0x000000479EC84F60> right = <__main__.Node object at 0x000000479ED05E80>
Create TreeLL
Value of Node =  TreeLL  left = None  right = None
Value  of  Node  =  Tree_Left  left  =  <__main__.Node  object  at
0x000000479ED0F160> right =  None
Value  of  Root  =  Root_Node   left  =  <__main__.Node  object  at
0x000000479EC84F60> right = <__main__.Node object at 0x000000479ED05E80>
```

3. 树的定义

树是一组存储元素的节点。这些节点具有父子关系。

（1）如果树非空，则有一个特殊节点称为树的根。根无父节点。

（2）树中每个与根不同的节点都有唯一的父节点。

12.3 树的列表表示

在列表中，将节点的值存储为第 1 个元素。第 2 个元素表示左子树的列表，第 3 个元素表示右子树的列表。图 12.5 显示了只有根节点的树。

图 12.5

在左边添加一个节点，如图 12.6 所示。

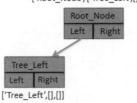

图 12.6

在右边再添加一个子节点，如图 12.7 所示。

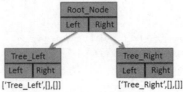

图 12.7

将节点添加到 Tree_Left 的左侧的方法如图 12.8 所示。

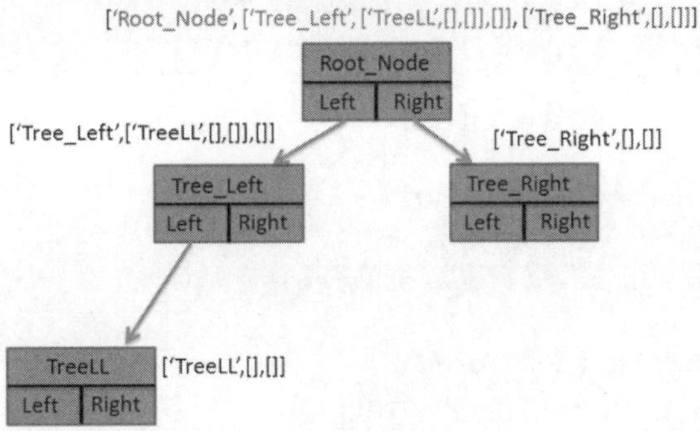

图 12.8

树可以用以下方式定义：

```
binary_tree = ['Root_Node',['Tree_Left',['TreeLL',[],[]],[]],['Tree_Right',[],[]]]
```

其中，根 Root_Node 在 binary_tree[0]；左子树在 binary_tree[1]；右子树在 binary_tree[2]。

现在编写代码。

步骤1：定义类。代码如下所示。

```
class Tree:
```

步骤2：创建构造函数。

在创建对象时，传递一个值。使用构造函数创建一个列表，这个值位于索引 0 处，而在索引 1 和 2 处有两个空列表。如果在左侧添加子树，则添加在索引 1 处，而对于右子树，将在索引 2 处插入。代码如下所示。

```
def __init__(self,data):
    self.tree = [data,[],[]]
```

步骤3：定义函数来插入左、右子树。

如果要在左子树中插入一个值，在索引 1 处弹出元素并在该位置插入新列表。类似地，在右侧插入子列表，则在索引 2 处弹出值并插入新列表。代码如下所示。

```
def left_subtree(self,branch):
    left_list = self.tree.pop(1)
    self.tree.insert(1,branch.tree)

def right_subtree(self,branch):
    right_list = self.tree.pop(2)
    self.tree.insert(2,branch.tree)
```

代码：

```
class Tree:
    def __init__(self,data):
        self.tree = [data,[],[]]

    def left_subtree(self,branch):
        left_list = self.tree.pop(1)
        self.tree.insert(1,branch.tree)

    def right_subtree(self,branch):
        right_list = self.tree.pop(2)
        self.tree.insert(2,branch.tree)
```

执行：

```
print("Create Root Node")
root = Tree("Root_node")
print("Value of Root = ",root.tree)
print("Create Left Tree")
tree_left = Tree("Tree_Left")
root.left_subtree(tree_left)
print("Value of Tree_Left = ",root.tree)
print("Create Right Tree")
tree_right = Tree("Tree_Right")
root.right_subtree(tree_right)
print("Value of Tree_Right = ",root.tree)
```

输出：

```
Create Root Node
Value of Root =  ['Root_node', [], []]
Create Left Tree
Value of Tree_Left =  ['Root_node', ['Tree_Left', [], []], []]
Create Right Tree
Value of Tree_Right =  ['Root_node', ['Tree_Left', [], []], ['Tree_Right', [], []]]
```

本段代码未考虑在两者之间插入 1 个子元素，插入子元素到指定位置，而原来该位置的子元素下推。

对插入函数进行更改。代码如下所示。

```
def left_subtree(self,branch):
    left_list = self.tree.pop(1)
    if len(left_list) > 1:
        branch.tree[1]=left_list
```

```
            self.tree.insert(1,branch.tree)
    else:
        self.tree.insert(1,branch.tree)
```

在左边插入子元素，首先弹出索引为 1 的元素。如果索引为 1 的元素的长度为 0，则只需插入列表；否则，将元素推到新子元素的左边。右子树也是如此。代码如下所示。

```
def right_subtree(self,branch):
    right_list = self.tree.pop(2)
    if len(right_list) > 1:
        branch.tree[2]=right_list
        self.tree.insert(2,branch.tree)
    else:
        self.tree.insert(2,branch.tree)

print("Create TreeLL")
treell = Tree("TreeLL")
tree_left.left_subtree(treell)
print("Value of Tree_Left = ",root.tree)
```

代码：

```
class Tree:
    def __init__(self,data):
        self.tree = [data,[],[]]

    def left_subtree(self,branch):
        left_list = self.tree.pop(1)
        if len(left_list) > 1:
            branch.tree[1]=left_list
            self.tree.insert(1,branch.tree)
        else:
            self.tree.insert(1,branch.tree)

    def right_subtree(self,branch):
        right_list = self.tree.pop(2)
        if len(right_list) > 1:
            branch.tree[2]=right_list
            self.tree.insert(2,branch.tree)
        else:
            self.tree.insert(2,branch.tree)
```

执行：

```
print("Create Root Node")
root = Tree("Root_node")
print("Value of Root = ",root.tree)
print("Create Left Tree")
tree_left = Tree("Tree_Left")
root.left_subtree(tree_left)
print("Value of Tree_Left = ",root.tree)
print("Create Right Tree")
tree_right = Tree("Tree_Right")
root.right_subtree(tree_right)
print("Value of Tree_Right = ",root.tree)
print("Create Left In between")
tree_inbtw = Tree("Tree left in between")
root.left_subtree(tree_inbtw)
print("Value of Tree_Left = ",root.tree)
print("Create TreeLL")
treell = Tree("TreeLL")
tree_left.left_subtree(treell)
print("Value of TREE = ",root.tree)
```

输出：

```
Create Root Node
Value of Root = ['Root_node',[],[]]
Create Left Tree
Value of Tree_Left = ['Root_node',['Tree_Left',[],[]],[]]
Create Right Tree
Value of Tree_Right = ['Root_node',['Tree_Left',[],[]],['Tree_
Right',[],[]]]
Create Left In between
Value of Tree_Left = ['Root_node',['Tree left in between',['Tree_
Left',[],[]],[]],
['Tree_Right',[],[]]]
Create TreeLL
Value of TREE = ['Root_node',['Tree left in between',['Tree_Left',
['TreeLL',[],[]],[]],[]],['Tree_Right',[],[]]]
```

下面介绍3种树的遍历方法。

(1) 先序遍历。

(2) 中序遍历。

(3) 后序遍历。

（1）先序遍历：首先访问根节点，然后访问节点上的所有节点，左边跟着右边的所有节点。

代码：

```python
class Node(object):
    def __init__(self, data_value):
        self.data_value = data_value
        self.left = None
        self.right = None
    def insert_left(self, child):
        if self.left is None:
            self.left = child
        else:
            child.left = self.left
            self.left = child
    def insert_right(self, child):
        if self.right is None:
            self.right = child
        else:
            child.right = self.right
            self.right = child
    def preorder(self, node):
        res = []
        if node:
            res.append(node.data_value)
            res = res + self.preorder(node.left)
            res = res + self.preorder(node.right)
        return res
```

执行：

```python
#Root_Node
print("Create Root Node")
root = Node("Root_Node")
#Tree_Left
print("Create Tree_Left")
tree_left = Node("Tree_Left")
root.insert_left(tree_left)
#Tree_Right
print("Create Tree_Right")
tree_right = Node("Tree_Right")
root.insert_right(tree_right)
```

```
#TreeLL
print("Create TreeLL")
treell = Node("TreeLL")
tree_left.insert_left(treell)
print("*****Preorder Traversal*****")
print(root.preorder(root))
```

输出：

```
Create Root Node
Create Tree_Left
Create Tree_Right
Create TreeLL
*****Preorder Traversal*****

['Root_Node','Tree_Left','TreeLL','Tree_Right']
>>>
```

(2) 中序遍历：首先访问左侧所有节点，最后访问根节点，然后访问右侧所有节点。

代码：

```
class Node(object):
    def __init__(self, data_value):
        self.data_value = data_value
        self.left = None
        self.right = None
    def insert_left(self, child):
        if self.left is None:
            self.left = child
        else:
            child.left = self.left
            self.left = child
    def insert_right(self, child):
        if self.right is None:
            self.right = child
        else:
            child.right = self.right
            self.right = child
    def inorder(self, node):
        res = []
        if node:
            res = self.inorder(node.left)
            res.append(node.data_value)
            res = res + self.inorder(node.right)
        return res
```

执行：

```
#Root_Node
print("Create Root Node")
root = Node("Root_Node")
#Tree_Left
print("Create Tree_Left")
tree_left = Node("Tree_Left")
root.insert_left(tree_left)
#Tree_Right
print("Create Tree_Right")
tree_right = Node("Tree_Right")
root.insert_right(tree_right)
#TreeLL
print("Create TreeLL")
treell = Node("TreeLL")
tree_left.insert_left(treell)
print("*****Inorder Traversal*****")
print(root.inorder(root))
```

输出：

```
Create Root Node
Create Tree_Left
Create Tree_Right
Create TreeLL
*****Inorder Traversal*****
['TreeLL','Tree_Left','Root_Node','Tree_Right']
>>>
```

（3）后序遍历：首先访问左侧所有节点，然后访问右侧所有节点，最后访问根节点。

代码：

```
class Node(object):
    def __init__(self, data_value):
        self.data_value = data_value
        self.left = None
        self.right = None
    def insert_left(self, child):
        if self.left is None:
            self.left = child
        else:
            child.left = self.left
            self.left = child
    def insert_right(self, child):
        if self.right is None:
            self.right = child
```

```python
            else:
                child.right = self.right
                self.right = child
    def postorder(self, node):
        res = []
        if node:
            res = self.postorder(node.left)
            res = res + self.postorder(node.right)
            res.append(node.data_value)
        return res
```

执行:

```python
#Root_Node
print("Create Root Node")
root = Node("Root_Node")
#Tree_Left
print("Create Tree_Left")
tree_left = Node("Tree_Left")
root.insert_left(tree_left)
#Tree_Right
print("Create Tree_Right")
tree_right = Node("Tree_Right")
root.insert_right(tree_right)
#TreeLL
print("Create TreeLL")
treell = Node("TreeLL")
tree_left.insert_left(treell)
print("*****Postorder Traversal*****")
print(root.postorder(root))
```

输出:

```
Create Root Node

Create Tree_Left
Create Tree_Right

Create Treell
*****Postorder Traversal*****

['TreeLL','Tree_Left','Tree_Right','Root_Node']
```

12.4 二叉堆

二叉堆是完整二叉树。除了最后一层，其余每层都被完全填补，表示树平衡。

每个新项都被插入可用空间的旁边。二叉堆可以存储在数组中。

二叉堆有两种类型——最小二叉堆和最大二叉堆，如图12.9所示。最小二叉堆：根节点是二叉堆所有节点中最小的，并且所有父节点都比其子节点小。最大二叉堆：根节点是二叉堆所有节点中最大的，并且所有父节点都比其子节点大。

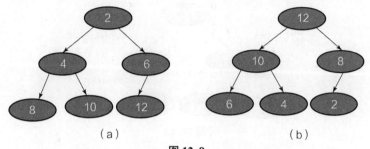

图 12.9

（a）最小二叉堆；（b）最大二叉堆

堆有以下两个重要属性。

（1）每层从左到右构造，最后一层可能未完全填满。每层按顺序填满。因此，插入值的顺序应为从左到右，逐层依次插入，如图12.10所示。

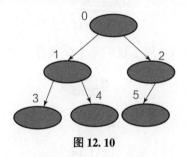

图 12.10

（2）最大二叉堆的父节点必须比其子节点大，最小二叉堆的父节点必须比其子节点小。图12.11所示是最大二叉堆的情况，所有父节点都大于其子节点。

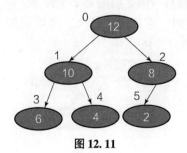

图 12.11

二叉堆可以用数组表示，如图12.12所示。

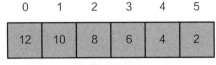

图 12.12

仔细观察该数组，会发现如果父节点存在位置n，左子节点就存在位置2n+1，右子节点存在位置2n+2，如图12.13所示。

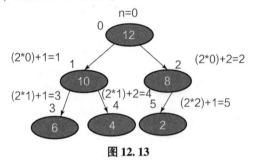

图 12.13

因此，如果知道父节点的位置，就很容易找到左、右子节点的位置。

假设使用以下值建立最大二叉堆：

$$20, 4, 90, 1, 125$$

步骤1：插入20，如图12.14所示。

图 12.14

步骤2：插入4→从左到右→第2层的第1个节点，如图12.15所示。

图 12.15

因为父节点大于子节点，所以继续操作。

步骤3：插入90→从左到右→第2层的第2个元素。

当插入90作为右子节点时，它违反了最大二叉堆规则，因为父节点的值为20。为解决该问题，交换位置，如图12.16所示。

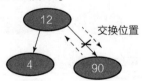

图 12.16

交换之后的最大二叉堆，如图 12.17 所示。

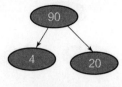

图 12.17

步骤 4：插入 1→从左到右→第 3 层的第 1 个元素，如图 12.18 所示。

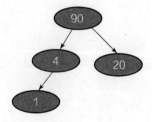

图 12.18

因为 1 比父节点小，所以继续操作。
步骤 5：插入 125。
因为违反了最大二叉堆规则，所以交换 4 和 125，如图 12.19 所示。

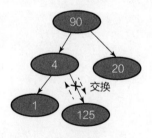

图 12.19

因为仍然不满足最大二叉堆规则，所以继续交换 125 和 90，如图 12.20 所示。

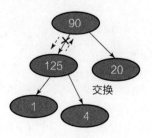

图 12.20

交换之后的堆如图 12.21 所示。

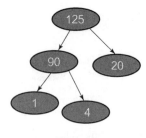

图 12.21

最大二叉堆构建成功。

示例 12.1

编写 Python 代码实现最大二叉堆。插入值，使根节点值最大，并且所有父节点都大于其子节点。

答：编写代码实现 MaxHeap 类。该类有两个函数。

（1） push()：插入值。

（2） float_up()：将值放置在其所属的位置。

步骤 1：定义 MaxHeap 类。代码如下所示。

```
class MaxHeap:
```

步骤 2：定义构造函数。代码如下所示。

```
def __init__(self):
    self.heap = []
```

步骤 3：定义 push() 函数。

push() 函数要完成两个工作。

（1） 将值追加到列表末尾（前面已看到，必须先在可用位置插入值）。

（2） 追加值后，push() 函数调用 float_up(index) 函数，将最后元素的索引值传递给 float_up() 函数，以便 float_up() 函数分析这些值并执行下一步操作。

代码如下所示。

```
def push(self,value):
    self.heap.append(value)
    self.float_up(len(self.heap) -1)
```

步骤 4：定义 float_up() 函数。

（1） float_up() 函数以索引值作为参数。代码如下所示。

```
def float_up(self,index):
```

（2）push()函数传递堆中最后元素的索引值。float_up()函数首先检查元素的索引值是否为0，如果是，则该元素是根节点。因为根节点无父节点，并且是堆的最后一个元素，同时无子节点，所以返回该值。代码如下所示。

```
if index == 0:
        return
```

（3）如果元素的索引值大于0，则继续操作。请看图12.22。

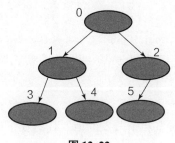

图 12.22

编程时还需要理解以下内容。

索引值为0的元素在位置1和2处有两个子节点。如果有元素在位置1，那么通过计算值（1// 2）的值找到父元素。类似地，如果有元素在位置2，通过计算（2// 2 – 1）的值找到父元素的插入值。因此，如果元素的索引值是奇数，那么其父元素的索引值定义为 parent_of_index = index// 2；如果元素的索引值是偶数，那么其父元素的索引值定义为 parent_of_index = (index// 2) – 1。这就成为 float_up()函数的外框架。右子节点的索引是偶数，左子节点的索引是奇数。

代码如下所示。

```
if index == 0:
        return
    else:
        if index % 2 == 0:
            parent_of_index = (index//2) - 1
            ------------ write code -------------
        else:
            parent_of_index = index//2
            ------------ write code -------------
```

（4）比较子节点和父节点的值。如果子节点的值大于父节点的值，则交换节点的值。代码如下所示。

```python
def float_up(self,index):
    if index ==0:
        return
    else:
        if index% 2 ==0:
            parent_of_index = (index//2)-1
            if self.heap[index] > self.heap[parent_of_index]:
                self.swap(index, parent_of_index)
        else:
            parent_of_index = index//2
            if self.heap[index] > self.heap[parent_of_index]:
                self.swap(index, parent_of_index)
        self.float_up(parent_of_index)
```

步骤 5：定义 swap( )函数。

使用 swap( )函数交换父节点和子节点的值。代码如下所示。

```python
def swap(self,index1, index2):
    temp = self.heap[index1]
    self.heap[index1] = self.heap[index2]
    self.heap[index2] = temp
```

代码：

```python
class MaxHeap:
    def __init__(self):
        self.heap = []
    def push(self,value):
        self.heap.append(value)
        self.float_up(len(self.heap)-1)
    def float_up(self,index):
        if index ==0:
            return
        else:
            if index%2 ==0:
                parent_of_index = (index//2)-1
                if self.heap[index] > self.heap[parent_of_index]:
                    self.swap(index, parent_of_index)
            else:
                parent_of_index = index//2
                if self.heap[index] > self.heap[parent_of_index]:
                    self.swap(index, parent_of_index)
            self.float_up(parent_of_index)
```

```python
        def peek(self):
            print(self.heap[0])
        def pop(self):
            if len(self.heap) >= 2:
                temp = self.heap[0]
                self.heap[0] = self.heap[len(self.heap) - 1]
                self.heap[len(self.heap) - 1]
                self.heap.pop()
                self.down_adj()
            elif len(self.heap) == 1:
                self.heap.pop()
            else:
                print("Nothing to pop")
        def swap(self, index1, index2):
            temp = self.heap[index1]
            self.heap[index1] = self.heap[index2]
            self.heap[index2] = temp
```

执行：

```python
H = MaxHeap()
print("*****pushing values*****")
print("pushing 165")
H.push(165)
print(H.heap)
print("pushing 60")
H.push(60)
print(H.heap)
print("pushing 179")
H.push(179)
print(H.heap)
print("pushing 400")
H.push(400)
print(H.heap)
print("pushing 6")
H.push(6)
print(H.heap)
print("pushing 275")
H.push(275)
print(H.heap)
```

输出：

```
*****pushing values*****
pushing 165
[165]
pushing 60
[165, 60]
pushing 179
[179, 60, 165]
pushing 400
[400, 179, 165, 60]
pushing 6
[400, 179, 165, 60, 6]
pushing 275
[400, 179, 275, 60, 6, 165]
>>>
```

示例 12.2

编写代码求最大二叉堆的最大值。

答：最大二叉堆的最大值很容易找到，因为最大值在堆的索引为 0 的根节点上。调用函数 peek()，将显示堆的最大值。代码如下所示。

```
def peek(self):
    print(self.heap[0])
```

示例 12.3

编写代码从最大二叉堆中弹出最大值。

答：包括以下两个步骤。

（1）将根节点与数组的最后一个元素交换并弹出值。

（2）根节点现在有最大值。因此，需要向下移动，将父节点与其左、右子节点进行比较，以确保子节点比父节点小。如果根节点没有最大值，将再次交换位置。

步骤1：定义 pop() 函数。

pop() 函数将根节点的值与列表的最后元素交换并弹出该值，然后调用 down_adj() 函数以向下移动并调整值。函数首先检查堆的大小。如果堆的长度是 1，则意味着只包含 1 个根节点，不需要进一步交换。代码如下所示。

```
def pop(self):
    if len(self.heap) >2:
        temp = self.heap[0]
        self.heap[0] = self.heap[len(self.heap) -1]
        self.heap[len(self.heap) -1]
        self.heap.pop()
        print("heap after popping largest value =", self.heap)
        self.down_adj()
```

```
            elif len(self.heap) ==1:
                self.heap.pop()
            else:
                print("Nothing to pop")
```

步骤2：定义 down_adj( ) 函数。

设置索引值为 0。

左子节点的索引 = left_child = index * 2 + 1。

右子节点的索引 = right_child = index * 2 + 2。

循环检查父节点和左、右子节点的值。如果父节点小于左子节点，则交换值。然后比较父节点与右子节点，如果父节点小于右子节点，则再次交换。

通过以下方式实现。

(1) 检查父节点是否小于左子节点。

①是，检查左子节点是否小于右子节点，

● 是，则将父节点与右子节点交换，

● 将索引值更改为右子节点的索引值。

②否，父节点和左子节点交换。

③将索引值设置为左子节点的索引值，以便进一步评价。

(2) 如果父节点不小于左子节点，但仅小于右子节点，则与右子节点交换值。

(3) 将索引值修改为右子节点的索引值。

代码如下所示。

```
def down_adj(self):
    index = 0
    for i in range(len(self.heap)//2):
        left_child = index*2 +1
        if left_child > len(self.heap):
            return
        print("left child = ", left_child)
        right_child = index*2 +2
        if right_child > len(self.heap):
            return
        print("right child = ", right_child)

        if self.heap[index] < self.heap[left_child]:
            temp = self.heap[index]
            self.heap[index] = self.heap[left_child]
            self.heap[left_child] = temp
            index = left_child
```

```
            if self.heap[index] < self.heap[right_child]:
                temp = self.heap[index]
                self.heap[index] = self.heap[right_child]
                self.heap[right_child] = temp
                index = right_child
```

代码:

```
class MaxHeap:
    def __init__(self):
        self.heap = []
    def push(self,value):
        self.heap.append(value)
        self.float_up(len(self.heap)-1)
    def float_up(self,index):
        if index==0:
            return
        else:
            if index%2==0:
                parent_of_index = (index//2)-1
                if self.heap[index] > self.heap[parent_of_index]:
                    temp = self.heap[parent_of_index]
                    self.heap[parent_of_index] = self.heap[index]
                    self.heap[index] = temp
            else:
                parent_of_index = index//2
                if self.heap[index] > self.heap[parent_of_index]:
                    temp = self.heap[parent_of_index]
                    self.heap[parent_of_index] = self.heap[index]
                    self.heap[index] = temp
            self.float_up(parent_of_index)
    def peek(self):
        print(self.heap[0])
    def pop(self):
        if len(self.heap)>=2:
            temp = self.heap[0]
            self.heap[0] = self.heap[len(self.heap)-1]
            self.heap[len(self.heap)-1]
            self.heap.pop()
            self.down_adj()
        elif len(self.heap)==1:
            self.heap.pop()
```

```python
        else:
            print("Nothing to pop")
    def swap(self,index1, index2):
        temp = self.heap[index1]
        self.heap[index1] = self.heap[index2]
        self.heap[index2] = temp
    def down_adj(self):
        index = 0

        for i in range(len(self.heap)//2):
            left_child = index*2 +1
            if left_child > len(self.heap) -1:
                print(self.heap)
                print("End Point")
                print("Heap value after pop() = ",self.heap)
                return
            right_child = index*2 +2
            if right_child > len(self.heap) -1:
                print("right child does not exist")
                if self.heap[index] < self.heap[left_child]:
                    self.swap(index,left_child)
                    index = left_child
                    print("Heap value after pop() = ",self.heap)
                return
            if self.heap[index] < self.heap[left_child]:
                if self.heap[left_child] < self.heap[right_child]:
                    self.swap(index,right_child)
                    index = right_child
                else:
                    self.swap(index,left_child)
                    index = left_child
            elif self.heap[index] < self.heap[right_child]:
                self.swap(index,right_child)
                index = right_child
            else:
                print("No change required" )
        print("Heap value after pop() = ",self.heap)
```

执行：

```
H = MaxHeap()
print("*****pushing values*****")
H.push(165)
print(H.heap)
H.push(60)
print(H.heap)
H.push(179)
print(H.heap)
H.push(400)
print(H.heap)
H.push(6)
print(H.heap)
H.push(275)
print(H.heap)
print("*****popping values*****")
H.pop()
H.pop()
H.pop()
H.pop()
H.pop()
H.pop()
H.pop()
```

输出：

```
pushing values
[165]
[165, 60]
[179, 60, 165]
[400, 179, 165, 60]
[400, 179, 165, 60, 6]
[400, 179, 275, 60, 6, 165]
*****popping values*****
[275, 179, 165, 60, 6]
End Point
Heap value after  pop() =  [275,179, 165, 60, 6]
right child does  not exist
Heap value after  pop() =  [179,60, 165, 6]
Heap value after  pop() =  [165,60, 6]
right child does not exist
Heap value after pop() =  [60, 6]
Heap value after pop() =  [6]
Nothing to pop
 >>>
```

示例 12.4

二叉堆的应用。

答：

（1）Dijkstra 算法。

（2）Prims 算法。

（3）优先队列。

（4）解决以下问题。

①获取数组中第 K 大的元素。

②排序几乎已有序的数组。

③合并 K 个已排序数组。

示例 12.5

何为优先队列？它如何实现？

答：

优先队列类似队列，但更高级。优先队列与队列有相同的方法。其与队列的主要区别：高优先级的值放在前面，最低优先级的值放在后面；元素从后面添加，从前面移除。优先级队列的元素按顺序添加，因此，每个元素都有优先级。优先级最高的元素首先被移除，优先级相同的元素按照队列顺序处理。

小结

第 11 章介绍了线性数据结构——链表。注意，树是层次化的数据结构，在树中搜索信息比在链表中搜索信息更快。

第 13 章

查找与排序

引言　　处理数据前按顺序对数据进行排序非常重要，这有助于简化搜索信息。本章介绍数据查找和排序的各种算法，以及这些算法的 Python 实现。

知识结构

- 顺序查找
- 对半查找
- 哈希排序
- 冒泡排序
- 选择排序
- 插入排序
- 希尔排序
- 快速排序

目标　　完成本章的学习后，读者应深入理解各种查找和排序算法的 Python 实现。

13.1　顺序查找

下面使用 Python 的成员运算符 in 检查列表中是否存在值。代码如下所示。

```
>>> list1 = [1,2,3,4,5,6,7]
>>> 3 in list1
True
>>> 8 in list1
False
>>>
```

本段代码实现在列表中查找元素，接下来探讨如何查找元素以及如何提高查找效率。这里从学习顺序查找开始。

在列表中查找元素的最简单方法是逐个检查。如果找到该元素，则查找结束，并返回该元素；否则，继续查找直到列表结束。这种查找方法称为线性查找或顺序查找，如图 13.1 所示。它遵循一个简单的方法，但这是查找元素的低效方法。因为如果元素不在列表中，将无法完成查找。

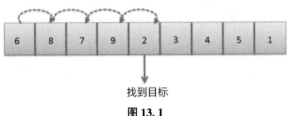

图 13.1

1. 顺序查找的实现

顺序查找按以下方式实现。

（1）函数用到两个参数：seq_list 表示列表；target_num 表示列表中要查找的数。

（2）如果在列表中找到 target_num，将 search_flag = 0 设置为 search_flag = 1；否则，将其设置为 0。

（3）循环访问列表，将列表的每个元素与 target_num 进行比较。

（4）找到匹配项，打印消息并将 search_flag 更新为 1。

（5）在 for 循环后，如果 search_flag 仍为 0，则表示未找到该数。

代码：

```
def sequential_search(seq_list, target_num):
    search_flag = 0
    for i in range(len(seq_list)):
        if seq_list[i] == target_num:
            print("Found the target number ", target_num, " at index",
            i,".")
            search_flag = 1;

    if search_flag == 0:
        print("Target Number Does Not Exist. Search Unsuccessful.")
```

执行：

```
seq_list = [1,2,3,4,5,6,7,8,2,9,10,11,12,13,14,15,16]
target_num = input("Please enter the target number : ")
sequential_search(seq_list, int(target_num))
```

输出1：

```
Please enter the target number : 5
Found the target number  5  at index 4 .
```

输出2：

```
Please enter the target number : 2
Found the target number  2  at index 1 .
Found the target number  2  at index 8 .
```

输出3：

```
Please enter the target number : 87
Target Number Does Not Exist. Search Unsuccessful.
```

2. 有序列表的顺序查找实现

列表元素排序后，多数不需要扫描整个列表。当查找到值大于目标数的元素时，不需要再进行查找。

步骤1：定义函数 sequential_search()，该函数用到两个参数——列表(seq_list)和目标数(target_num)。代码如下所示。

```
def sequential_search(seq_list,target_num):
```

步骤2：首先设置定义标志（search_flag）并将其设置为 False 或 0。如果找到该元素，则将标志设置为 True 或 1。因此，遍历列表后，如果 search_flag 值为 False 或 0，则表明列表中不存在该数。代码如下所示。

```
def sequential_search(seq_list,target_num):
    search_flag = 0
```

步骤3：定义 for 循环逐个判断元素。代码如下所示。

```
def sequential_search(seq_list,target_num):
    search_flag = 0
    for i in range(len(seq_list)):
```

步骤4：定义元素比较。因为对于有序列表中的每个 i，必须检查是否 i > target_num。如果是，则表示达到一个大于目标数的元素。但是，如果 seq_list[i] == target_num，则表示查找成功，将 search_flag 设置为 1。代码如下所示。

```
def sequential_search(seq_list,target_num):
    search_flag = 0
    for i in range(len(seq_list)):
```

```
            if seq_list[i] > target_num:
                print("search no further.")
                break;
            elif seq_list[i] == target_num:
                print("Found the target number ", target_num, " at index",
                    i,".")
                search_flag = 1
```

步骤5：执行 for 循环后，如果 search_flag 的值仍为 0，则显示消息指出未找到目标数。

代码：

```
def sequential_search(seq_list, target_num):
    search_flag = 0
    for i in range(len(seq_list)):
        if seq_list[i] > target_num:
            print("search no further.")
            break;
        elif seq_list[i] == target_num:
            print("Found the target number ", target_num, " at index",
                i,".")
            search_flag = 1

    if search_flag == 0:
        print("Target Number Does Not Exist. Search Unsuccessful.")
```

执行：

```
seq_list = [1,2,2,3,4,5,6,7,8,9,10,11,12,13,14,15,16]
target_num = input("Please enter the target number : ")
sequential_search(seq_list, int(target_num))
```

输出1：

```
Please enter the target number : 2
Found the target number  2  at index 1 .
Found the target number  2  at index 2 .
search no further.
>>>
```

输出2：

```
Please enter the target number : 8
Found the target number  8  at index 8 .
search no further.
>>>
```

输出 3：

```
Please enter the target number：89
Target Number Does Not Exist.Search Unsuccessful.
>>>
```

13.2　对半查找

对半查找用于从已排序列表中查找目标值，如图 13.2 所示。从列表的中心开始查找，比较中心元素和目标值。如果目标值大于中心元素，则在列表的右半部分查找目标值，并不需要考虑左半部分。同样，如果目标值小于中心元素，则在列表的左半部分查找目标值。重复此过程，直到查找完成。对半查找的优点在于，在每个查找操作中，列表都被分成两半，焦点仅转移到可能性较大的那一半。

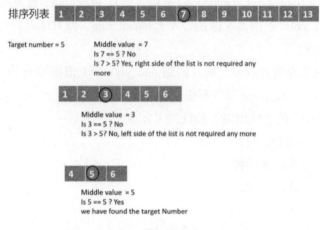

图 13.2

对半查找按以下方式实现。

步骤 1：定义 binary_search( ) 函数。需要以下 4 个参数。

（1）sorted_list：排序形式的输入列表。

（2）target_num：目标值。

（3）start_point：查找起点，默认 start_point = 0。

（4）end_point：查找终点，默认 end_point = None。

注意，在每个步骤中将列表分成两半，因此查找起点和查找终点都可能发生变化。

```
def binary_search(sorted_list, target_num, start_point = 0, end_point = None):
```

步骤 2：执行以下操作。

（1）将 search_flag 设置为 False。

（2）如果未提供 end_point，则默认值为 None，将其设置为输入列表的长度。

代码如下所示。

```
def binary_search(sorted_list, target_num, start_point = 0, end_point = None):
    search_flag = False
    if end_point == None:
        end_point = len(sorted_list) -1
```

步骤 3：检查，start_point 应小于 end_point。如果为真，则执行以下操作。

（1）读取中间索引值：mid_point =（end_point + start_point）//2。

（2）判断 mid_point 处的元素值是否等于 target_num。

如果 sorted_list[mid_point] == target_num，则将 search_flag 设置为 True。

（3）判断 mid_point 处的元素值是否大于 target_num，即判断是否有 sorted_list[mid_point] > target_num。

①如果是，则丢弃列表右侧，从开始到 mid_point - 1 值重复查找。将 end_point 设置为 mid_point - 1，start_point 保持不变（0）。

②binary_search() 函数使用以下参数调用自身。

- sorted_list：和以前一样。
- target_num：和以前一样。
- start_point：和以前一样。
- end_point：mid_point - 1。

③如果不是，判断 mid_point 处的元素值是否小于 target_num，即判断是否有 sorted_list[mid_point] < target_num。

- 如果是，则丢弃列表左侧。从 mid_point + 1 到列表末尾重复查找。start_point 设置为 mid_point +1，end_point 保持不变。
- binary_search() 函数使用以下参数调用自身。
    ◆ sorted_list：和以前一样。
    ◆ target_num：和以前一样。
    ◆ start_point：mid_point +1。
    ◆ end_point：和以前一样。

（4）如果按此过程结束时 search_flag 仍为 False，则表明列表无该值。

代码：

```
def binary_search(sorted_list, target_num, start_point = 0, end_point = None):
    search_flag = False
    if end_point == None:
        end_point = len(sorted_list) -1
    if start_point < end_point:
        mid_point = (end_point + start_point)//2
        if sorted_list[mid_point] == target_num:
            search_flag = True
            print(target_num," Exists in the list at ",sorted_list.index(target_num))
        elif sorted_list[mid_point] > target_num:
            end_point = mid_point -1
            binary_search(sorted_list, target_num,start_point, end_point)
        elif sorted_list[mid_point] < target_num:
            start_point = mid_point +1
            binary_search(sorted_list, target_num, start_point, end_point)
    elif not search_flag:
        print(target_num," Value does not exist")
```

执行：

```
sorted_list =[1,2,3,4,5,6,7,8,9,10,11,12,13]
binary_search(sorted_list,14)
binary_search(sorted_list,0)
binary_search(sorted_list,5)
```

输出：

```
14  Value does not exist
0   Value does not exist
5   Exists in the list at  4
```

注意：以下内容涉及时间复杂度。如果不理解 Big – O 表示，请参阅本书附录部分。

13.3 哈希排序

哈希表是使用哈希函数为数据元素生成索引或地址值的数据结构，用于实现关联数组，该数组将键映射到值。其好处是，允许更快访问数据，因为索引值充当数值的键。哈希表用键–值对存储数据，但数据使用哈希函数生成。Python 中的哈希表为字典数据类型。字典的键使用哈希函数生成，字典的数据元素的顺序不固定。读者已经理解用于访问字典对象的各种函数，但实际上要学习哈希表的实现方式。二叉搜索树实现各种操作的时间复杂度为 O(logn)。如何让搜索操作更快？能否达到 O(1) 的时间复杂度？这正是哈希表产生的原因。与已知索引的列表或数组一样，搜索操作的时间复杂度可能会变为 O(1)。类似地，如果数据存储在键-值对中，则能更快查找结果。为此，有键，有放置值的位置。如果能在位置和键之间建立关系，就更容易实现较快速度的检索，如图 13.3 所示。

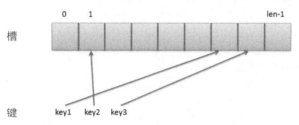

图 13.3

因为键、值并不总是非负整数，也可能是字符串，所以使用预哈希将字符串键与索引匹配。对于每个键，都需要在数组中找到索引，在该索引对应位置放置相应的值。为此，需要创建 hash( ) 函数，该函数可以将任何类型的键映射到随机数组索引。在此过程中可能出现冲突，即映射两个键指向同一个索引，如图 13.4 所示。

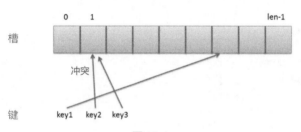

图 13.4

为了解决冲突，需要使用链表，如图 13.5 所示。

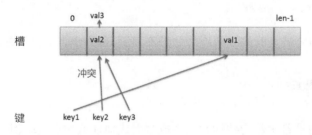

图 13.5

但是，对于同一个位置，可能存在多个冲突，考虑到需要将所有值作为链表元素插入的最坏情况，这将对时间复杂度产生严重影响。在最坏的情况下，将所有值作为链表元素放在同一个索引中。为了避免这种情况，考虑开放式寻址过程。开放式寻址是创建新地址的过程。考虑这样一种情况：如果存在冲突，则将索引递增1并将值放置在该索引处，如图13.6所示，放置val3时存在冲突，因为val2已存在于索引1中。因此，索引值递增1(1+1=2)，将val3存放于索引2中。

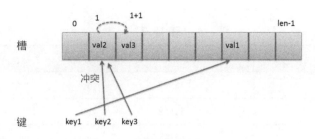

图13.6

如果在索引2处有其他值，则索引再次递增，val3可放在索引3处。重复递增索引过程直到发现空位置为止，称为线性探测。另外，二次探测以索引值的2倍递增。因此，空位置查找在1，2，4，8的距离上完成，依此类推。重新哈希是再次对获得结果进行哈希处理以查找空位置的过程。

哈希函数的目的是计算一个索引，从中可以找到正确的值。具体工作流程如下。

(1) 在数组中统一分布键。

(2) 如果n是键的数量，m是数组大小，则在使用整数作为键的情况下，hash( ) = n%m（模运算符）。

①对于数组和哈希函数，更倾向于使用质数来实现均匀分布。

②对于字符串键，可以计算每个字符的ASCII值，将其相加并对其进行模运算。

多数情况下，哈希表证明比搜索树更有效，通常用于缓存、数据库和集。

哈希函数的要点如下。

(1) 使用质数避免聚类。

(2) 条目数除以数组大小称为负载因子。

(3) 如果负载因子增加，则冲突次数增加导致哈希表性能降低。

(4) 负载因子超过给定阈值时要调整表的大小。但是，这将是代价大的选项，因为调整表的大小时，输入值的哈希值都会更改，需要O(n)才能完成。因此，动态数组不适合实时方案。

哈希函数的目的是将值或条目映射到哈希表中的可用位置。因此，对于每个条目，哈希函数计算一个整数值，该值在0~(m-1)的范围内，其中m是数组长度。

1. 余数哈希函数

余数哈希函数通过从集合中一次读取一项的方式计算索引值，然后将其除以数组大小，并返回其余数部分。

$$h(item) = item \% m, m = 数组大小$$

分析数组：[18，12，45，34，89，4]。

上面数组的大小为 8。

缺点：从表 13.1 中可以看到，18 和 34 具有相同哈希值 2，12 和 4 具有相同哈希值 4。这是执行程序时发生冲突的结果，将值 18 和 12 替换为 34 和 4，哈希表中找不到这些值。

表 13.1

item	Calculation = item%m	结果
18	18%8	2
12	12%8	4
45	45%8	5
34	34%8	2
89	89%8	1
4	4%8	4

函数的实现过程如下。

步骤 1：定义 hash() 函数。代码如下所示。

```
def hash(list_items, size):
```

步骤 2：执行以下步骤。

（1）创建空列表。

（2）将此键从数字 0 填充到 size。此示例中的列表有 8 个元素，因此创建列表 [0, 1, 2, 3, 4, 5, 6, 7]。

（3）使用 fromkeys() 函数将列表转换为 dict，得到字典对象的形式 {0: None, 1: None, 2: None, 3: None, 4: None, 5: None, 6: None, 7: None}，将此值赋值给 hash_table。代码如下所示。

```
def hash(list_items, size):
    temp_list = []
    for i in range(size):
        temp_list.append(i)
    hash_table = dict.fromkeys(temp_list)
```

步骤 3：

（1）循环访问列表。

（2）计算每项的索引值 item%size。

（3）对于 hash_table = index 中的键、值，请插入该项。

代码：

```
def hash(list_items, size):
    temp_list =[]
    for i in range(size):
        temp_list.append(i)
    print(temp_list)
    hash_table = dict.fromkeys(temp_list)
    print(hash_table)
    for item in list_items:
        i = item%size
        hash_table[i] = item
    print("value of hash table is : ",hash_table)
```

执行：

```
list_items = [18,12,45,34,89,4]
hash(list_items, 8)
```

输出：

```
[0, 1, 2, 3, 4, 5, 6, 7]
{0: None, 1: None, 2: None, 3: None, 4: None, 5: None, 6: None, 7: None}
value of hash table is :{0: None, 1: 89, 2: 34, 3: None, 4: 4, 5: 45, 6: None, 7: None}
>>>
```

2. 折叠哈希函数

折叠哈希函数是用于避免哈希产生冲突的技术。项被分成大小相等的片段并加在一起，然后使用相同哈希函数（item% size）计算位置。

假设给定一个电话列表：

```
phone_list =[4567774321,4567775514,9851742433,4368884732]
```

将每个数转换为字符串，然后将每个字符串转换为列表，最后将每个列表加到另一个列表中，得到以下结果：

```
[['4','5','6','7','7','7','4','3','2','1'],['4','5','6','7','7','7','5','5','1','4'],
['9','8','5','1','7','4','2','4','3','3'],['4','3','6','8','8','8','4','7','3','2']]
```

从新列表中逐个读取列表项，对于每项，连接两个字符，将其转换为整数，然后连接旁边的字符，将其转换为整数并添加两个值，继续此操作，直到添加所有元素。计算结果如下所示。

```
ssss['4','5','6','7','7','7','4','3','2','1']
```

(1) items = 4 5
　　string val = 45
　　integer value = 45
　　hash value = 45

(2) items = 6 7
　　string val = 67
　　integer value = 67
　　hash value = 45 + 67 = 112

(3) items = 7 7
　　string val = 77
　　integer value = 77
　　hash value = 112 + 77 = 189

(4) items = 4 3
　　string val = 43
　　integer value = 43
　　hash value = 189 + 43 = 232

(5) items = 2 1
　　string val = 21
　　integer value = 21
　　hash value = 232 + 21 = 253

同样地，有如下结果。

['4', '5', '6', '7', '7', '7', '5', '5', '1', '4'] 的哈希值为 511。
['9', '8', '5', '1', '7', '4', '2', '4', '3', '3'] 的哈希值为 791。
['4', '3', '6', '8', '8', '8', '4', '7', '3', '2'] 的哈希值为 1069。

现在，当数组大小为 11 时，调用哈希函数计算 [253, 511, 791, 1069] 的哈希值，如表 13.2 所示。

表 13.2

item	Calculation = item % m	结果
253	253 % 11	0
511	511 % 11	5
791	791 % 11	10
1069	1069 % 11	2

因此，输出结果如下所示。

{0:253, 1: None, 2:1069, 3: None, 4: None, 5: 511, 6: None, 7: None, 8: None, 9: None, 10:791}

3. 实现

查看程序执行语句：

```
phone_list = [4567774321, 4567775514, 9851742433, 4368884732]
str_phone_values = convert_to_string(phone_list)
folded_value = folding_hash(str_phone_values)
folding_hash_table = hash(folded_value,11)
print(folding_hash_table)
```

（1）定义电话号码列表：phone_list = [4567774321, 4567775514, 9851742433, 4368884732]。

（2）语句"str_phone_values = convert_to_string(phone_list)"调用函数 convert_to_string()，并将 phone_list 作为参数传递。该函数返回列表的列表，每次读取电话号码将其转换为列表并添加到新列表中。得到结果：[['4', '5', '6', '7', '7', '7', '4', '3', '2', '1'], ['4', '5', '6', '7', '7', '7', '5', '5', '1', '4'], ['9', '8', '5', '1', '7', '4', '2', '4', '3', '3'], ['4', '3', '6', '8', '8', '8', '4', '7', '3', '2']]。

函数的实现过程如下。

（1）定义两个列表：phone_list = [ ]。

（2）对于 phone_list 中的元素，逐个读取电话号码。

①将电话号码转换为字符串：temp_string = str(i)。

②将每个字符串转换为列表：temp_list = list(temp_string)。

③将获得的列表追加到步骤（1）中定义的 phone_list。

④返回 phone_list 并将值赋给 str_phone_values。

代码如下所示。

```
def convert_to_string(input_list):
    phone_list =[]
    for i in input_list:
        temp_string = str(i)
        temp_list = list(temp_string)
        phone_list.append(temp_list)
    return phone_list
```

（3）将列表 str_phone_values 传递给函数 folding_hash()。该函数将列表作为输入。

①访问 phone_list 中的每个元素，该元素也是列表。

②逐个读取列表项。

③将每项的前两个字符连接并将其转换为整数,然后连接旁边的字符,将其转换为整数,并将两个值相加。
④从列表中弹出前两个元素。
⑤重复③和④,直到添加所有元素。
⑥函数返回哈希值的列表。

代码如下所示。

```python
def folding_hash(input_list):
    hash_final = []
    while len(input_list) > 0:
        hash_val = 0
        for element in input_list:
            while len(element) > 1:
                string1 = element[0]
                string2 = element[1]
                str_combine = string1 + string2
                int_combine = int(str_combine)
                hash_val += int_combine
                element.pop(0)
                element.pop(0)
            if len(element) > 0:
                hash_val += element[0]

            else:
                pass
        hash_final.append(hash_val)
    return hash_final
```

(4) 调用哈希函数。哈希函数代码相同。代码如下所示。

```python
def hash(list_items, size):
    temp_list = []
    for i in range(size):
        temp_list.append(i)
    hash_table = dict.fromkeys(temp_list)
    for item in list_items:
        i = item % size
        hash_table[i] = item
    return hash_table
```

代码：

```python
def hash(list_items, size):
    temp_list =[]
    for i in range(size):
        temp_list.append(i)
    hash_table = dict.fromkeys(temp_list)
    for item in list_items:
        i = item%size
        hash_table[i] = item
    return hash_table
def convert_to_string(input_list):
    phone_list =[]
    for i in input_list:
        temp_string = str(i)
        temp_list = list(temp_string)
        phone_list.append(temp_list)
    return phone_list
def folding_hash(input_list):
    hash_final = []
    while len(input_list) > 0:
        hash_val = 0
        for element in input_list:
            while len(element) > 1:
                string1 = element[0]
                string2 = element[1]
                str_combine = string1 + string2
                int_combine = int(str_combine)
                hash_val += int_combine
                element.pop(0)
                element.pop(0)
            if len(element) > 0:
                hash_val += element[0]
            else:
                pass
            hash_final.append(hash_val)
        return hash_final
```

执行：

```python
phone_list = [4567774321, 4567775514, 9851742433, 4368884732]
str_phone_values = convert_to_string(phone_list)
folded_value = folding_hash(str_phone_values)
folding_hash_table = hash(folded_value,11)
print(folding_hash_table)
```

输出:

{0: 253, 1: None, 2: 1069, 3: None, 4: None, 5: 511, 6: None, 7: None, 8: None, 9: None, 10: 791}

为了在索引中存储电话号码,稍微更改了 hash( )函数。
(1) hash( )函数使用 1 个参数:phone_list。
(2) 计算索引后,从 phone_list 保存而不是 folded_value。
代码如下所示。

```python
def hash(list_items,phone_list, size):
    temp_list =[]
    for i in range(size):
        temp_list.append(i)
    hash_table = dict.fromkeys(temp_list)
    for i in range(len(list_items)):
        hash_index = list_items[i]%size
        hash_table[hash_index] = phone_list[i]
    return hash_table
```

最终代码:

```python
def hash(list_items,phone_list, size):
    temp_list =[]
    for i in range(size):
        temp_list.append(i)
    hash_table = dict.fromkeys(temp_list)
    for i in range(len(list_items)):
        hash_index = list_items[i]% size
        hash_table[hash_index] = phone_list[i]
    return hash_table

def convert_to_string(input_list):
    phone_list =[]
    for i in input_list:
        temp_string = str(i)
        temp_list = list(temp_string)
        phone_list.append(temp_list)
    return phone_list

def folding_hash(input_list):
    hash_final = []
```

```
        while len(input_list) > 0:
            hash_val = 0
            for element in input_list:
                while len(element) > 1:
                    string1 = element[0]
                    string2 = element[1]
                    str_combine = string1 + string2
                    int_combine = int(str_combine)
                    hash_val += int_combine
                    element.pop(0)
                    element.pop(0)
                if len(element) > 0:
                    hash_val += int(element[0])
                else:
                    pass
                hash_final.append(hash_val)
        return hash_final
```

执行：

```
phone_list = [4567774321, 4567775514, 9851742433, 4368884732]
str_phone_values = convert_to_string(phone_list)
folded_value = folding_hash(str_phone_values)
folding_hash_table = hash(folded_value,phone_list,11)
print(folding_hash_table)
```

输出：

```
{0: 4567774321, 1: None, 2: 4368884732, 3: None, 4: None, 5: 4567775514, 6: None,
7: None, 8: None, 9: None, 10: 9851742433}
```

13.4 冒泡排序

冒泡排序也称为下沉排序或比较排序，如图 13.7 所示。在冒泡排序中，每个元素都与相邻元素比较，如果顺序错误，则交换元素。这种算法非常耗时，虽然简单，但效率很低。

实现冒泡排序非常简单。

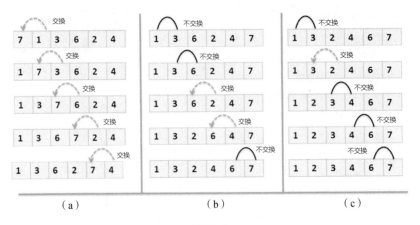

图 13.7

步骤 1：定义 bubble_sort( ) 函数。将排序列表作为输入。代码如下所示。

```
def bubble_sort(input_list):
```

步骤 2：设置 for i in range(len(input_list)) 循环。

（1）在 for 循环中，设置另一个 for j in range(len(input_list) - i - 1) 循环。

（2）对于每个 i，在索引 j 处的嵌套循环值与索引 j+1 处的值进行比较。如果索引 j+1 处的值小于索引 j 处的值，则交换值。

（3）for 循环结束后，输出已排序列表。

代码：

```
def bubble_sort(input_list):
    for i in range(len(input_list)):
        for j in range(len(input_list) - i - 1):
            if input_list[j] > input_list[j+1]:
                temp = input_list[j]
                input_list[j] = input_list[j+1]
                input_list[j+1] = temp
    print(input_list)
```

执行：

```
x = [7,1,3,6,2,4]
print("Executing Bubble sort for ",x)
bubble_sort(x)
y = [23,67,12,3,45,87,98,34]
print("Executing Bubble sort for ",y)
bubble_sort(y)
```

输出：

```
Executing Bubble sort for [7,1,3,6,2,4]
[1,2,3,4,6,7]
Executing Bubble sort for [23,67,12,3,45,87,98,34]
[3,12,23,34,45,67,87,98]
```

13.5 选择排序

选择排序的实现过程如下。

步骤1：定义 selection_sort( ) 函数。需要排序列表作为输入。代码如下所示。

```
def selection_sort(input_list):
```

步骤2：设置 for i in range(len(input_list) – 1) 循环。

（1）在 for 循环中，设置另一个循环 for j in range(i + 1,len(input_ list))。

（2）对于每个 i，索引 j 处的嵌套循环值与索引 i 处的值进行比较。如果索引 j 的值小于索引 i 的值，则交换值。

（3）for 循环结束后，打印已排序列表。

代码：

```
def selection_sort(input_list):
    for i in range(len(input_list)-1):
        for j in range(i+1,len(input_list)):
            if input_list[j] < input_list[i]:
                temp = input_list[j]
                input_list[j] = input_list[i]
                input_list[i] = temp
    print(input_list)
```

执行：

```
selection_sort([15,10,3,19,80,75])
selection_sort([5,9,80,65,71,24,15,10,3,19,85,75])
```

输出：

```
[3,10,15,19,75,80]
[3,5,9,10,15,19,24,65,71,75,80,85]
>>>
```

13.6　插入排序

在插入排序中，将位置 x 处的元素与位置 x − 1 至位置 0 处的每个元素进行比较。如果发现元素小于与之比较的任何值，则交换值。重复此过程，直到比较完最后一个元素，如图 13.8 所示。

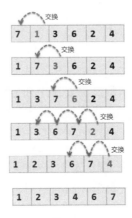

图 13.8

实现插入排序非常容易，如下所示。

分析列表 [7, 1, 3, 6, 2, 4]，令 indexi = i, indexj = indexi + 1，如表 13.3 所示。

表 13.3

indexi	indexj	val[i] < val[j]	是否交换	改变索引值
0	1	1 < 7	是	indexi = indexi − 1 = −1 indexj = indexj − 1 = 0
Iteration 2	List：1, 7, 3, 6, 2, 4			
1	2	3 < 7	是	indexi = indexi − 1 = 0 indexj = indexj − 1 = 1
	List：1, 3, 7, 6, 2, 4			
0	1	3 < 1	否	indexi = indexi − 1 = −1
Iteration 3	List：1, 3, 7, 6, 2, 4			
2	3	6 < 7	是	indexi = indexi − 1 = 1 indexj = indexj − 1 = 0

续表

indexi	indexj	val［i］< val［j］	是否交换	改变索引值
	List：1, 3, 6, 7, 2, 4			
1	2	7 < 3	否	indexi = indexi − 1 = 0
0	2	7 < 1	否	indexi = indexi − 1 = −1
Iteration 4	List：1, 3, 6, 7, 2, 4			
3	4	2 < 7	是	indexi = indexi − 1 = 2 indexj = indexj − 1 = 1
	List：1, 3, 6, 2, 7, 4			
2	3	2 < 6	是	indexi = indexi − 1 = 1 indexj = indexj − 1 = 2
	List：1, 3, 2, 6, 7, 4			
1	2	2 < 3	是	indexi = indexi − 1 = 0 indexj = indexj − 1 = 1
	List：1, 2, 3, 6, 7, 4			
0	1	2 < 1	否	indexi = indexi − 1 = −1
Iteration 5	List：1, 2, 3, 6, 7, 4			
4	5	4 < 7	是	indexi = indexi − 1 = 3 indexj = indexj − 1 = 4
	List：1, 2, 3, 6, 4, 7			
3	4	4 < 6	是	indexi = indexi − 1 = 2 indexj = indexj − 1 = 1
	List：1, 2, 3, 4, 6, 7			
2	3	4 < 3	否	indexi = indexi − 1 = 1
1	3	4 < 2	否	indexi = indexi − 1 = 0
0	3	4 < 1	否	indexi = indexi − 1 = −1

步骤1：定义 insertion_sort( ) 函数，将列表 input_list 作为输入。代码如下所示。

```
def insertion_sort(input_list):
```

步骤 2：使用语句 for i in range(len(input_list)-1)，设置 indexi = i, indexj = i+1。代码如下所示。

```
for i in range(len(input_list) -1):
    indexi = i
    indexj = i +1
```

步骤 3：设置 while 循环，条件为 indexi >=0。

如果 input_list[indexi] > input_list[indexj]，则交换 input_list[indexi] 和 input_list[indexj]，令 indexi = indexi -1, indexj = indexj -1；否则，令 indexi = indexi -1。代码如下所示。

```
while indexi >= 0:
    if input_list[indexi] > input_list[indexj]:
        print("swapping")
        temp = input_list[indexi]
        input_list[indexi] = input_list[indexj]
        input_list[indexj] = temp
        indexi = indexi - 1
        indexj = indexj - 1
    else:
        indexi = indexi - 1
```

步骤 4：打印更新列表。

代码：

```
def insertion_sort(input_list):
    for i in range(len(input_list) -1):
        indexi = i
        indexj = i +1
        print("indexi = ", indexi)
        print("indexj = ", indexj)
        while indexi >=0:
            if input_list[indexi] > input_list[indexj]:
                print("swapping")
                temp = input_list[indexi]
                input_list[indexi] = input_list[indexj]
                input_list[indexj] = temp
                indexi = indexi - 1
                indexj = indexj - 1
            else:
                indexi = indexi - 1
        print("list update:",input_list)
    print("final list = ", input_list)
```

执行：

```
insertion_sort([9,5,4,6,7,8,2])
```

输出：

```
indexi = 0
indexj = 1
swapping
list update:[5,9,4,6,7,8,2]
indexi = 1
indexj = 2
swapping
swapping
list update:[4,5,9,6,7,8,2]
indexi = 2
indexj = 3
swapping
list update:[4,5,6,9,7,8,2]
indexi = 3
indexj = 4
swapping
list update:[4,5,6,7,9,8,2]
indexi = 4
indexj = 5
swapping
list update:[4,5,6,7,8,9,2]
indexi = 5
indexj = 6
swapping
swapping
swapping
swapping
swapping
swapping
list update: [2,4,5,6,7,8,9]
final list = [2,4,5,6,7,8,9]
```

13.7 希尔排序

希尔排序是插入排序的一种，又称为"缩小增量排序"。它具有以下特点。

(1)非常高效。
(2)基于插入排序。
(3)首先在广泛分布元素上实现插入排序,然后每进行一步,空间或间隔都会缩小。
(4)非常适合中等大小的数据集。
(5)最坏情况时间复杂度:O(n)。
(6)最坏情况空间复杂度:O(n)。

希尔排序的算法如下。

(1)分析列表[10,30,11,4,36,31,15,1],列表大小 n = 8,n/2 = 4,将该值命名为 k,如图 13.9 所示。

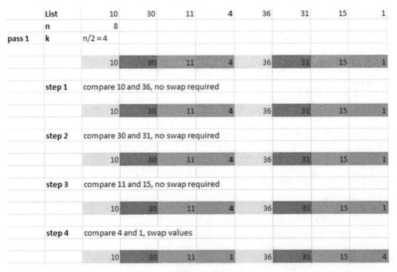

图 13.9

(2)令 k = k/2 = 4/2 = 2,分析每个第 k 个元素并对其顺序排序,如图 13.10 所示。

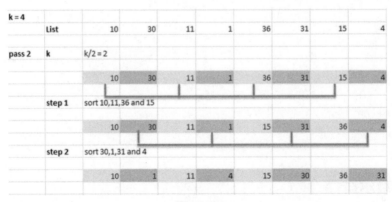

图 13.10

(3)令 k = k/2 = 2/2 = 1,这是最后通道,始终是插入通道,如图 13.11 所示。

	k = 2								
		List	10	30	11	1	36	31	15	4
pass 3	k	k/2 = 1								
			10	30	11	1	36	31	15	4
		step 1	1	4	10	11	15	30	31	36

* The last Pass is always an insertion Pass..

图 13.11

下面介绍希尔排序的实现。

步骤 1：定义 shell_sort( ) 函数对列表排序。使用列表 input_list 作为输入值排序。代码如下所示。

```
def shell_sort(input_list):
```

步骤 2：计算列表的大小，n = len(input_list)，while 循环步数 k = n/2。代码如下所示。

```
def shell_sort(input_list):
    n = len(input_list)
    k = n//2
```

步骤 3：排序。

当 k > 0 时，输入列表大小。如果列表在索引 i 处的值小于位于索引 i - k 处的值，则交换两个值，最后令 k = k//2。代码如下所示。

```
while k > 0:
    for j in range(n):
        for i in range(k,n):
            temp = input_list[i]
            if input_list[i] < input_list[i-k]:
                input_list[i] = input_list[i-k]
                input_list[i-k] = temp
    k = k//2
```

步骤 4：打印已排序列表的值。

代码：

```
def shell_sort(input_list):
    #计算列表 input_list 的长度
    n = len(input_list)
    #计算 k = n/2
    k = n//2

    #创建循环进行排序
```

```
        while k > 0:
            for j in range(n):
                for i in range(k,n):
                    temp = input_list[i]
                    if input_list[i] < input_list[i-k]:
                        input_list[i] = input_list[i-k]
                        input_list[i-k] = temp
            k = k//2
    print(input_list)
```

执行：

```
shell_sort([10,9,89,30,11,1,36,31,15,4])
```

输出：

```
[1, 4, 9, 10, 11, 15, 30, 31, 36, 89]
```

13.8 快速排序

快速排序在每一轮中使用中心点（pivot）比较数，将所有小于中心点的项都移到列表左侧，将所有大于中心点的项都移到列表右侧，得到所有值都小于中心点的左分区和所有值都大于中心点的右分区。

快速排序算法如下。

（1）分析列表 [15, 39, 4, 20, 50, 6, 28, 2, 13]，将最后一个元素 13 视为中心点，将第一个元素 15 作为左标记（leftmark），将倒数第二个元素 2 作为右标记（rightmark），如图 13.12 所示。

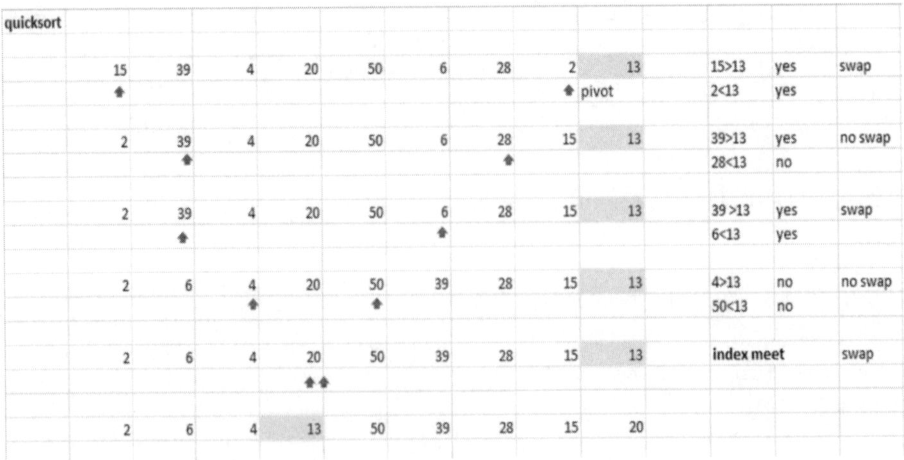

quicksort											
	15	39	4	20	50	6	28	2	13	15>13	yes	swap
	↑								↑ pivot	2<13	yes	
	2	39	4	20	50	6	28	15	13	39>13	yes	no swap
		↑					↑			28<13	no	
	2	39	4	20	50	6	28	15	13	39>13	yes	swap
		↑				↑				6<13	yes	
	2	6	4	20	50	39	28	15	13	4>13	no	no swap
			↑		↑					50<13	no	
	2	6	4	20	50	39	28	15	13	index meet		swap
				↑↑								
	2	6	4	13	50	39	28	15	20			

图 13.12

①如果 leftmark > pivot 和 rightmark < pivot，则交换 leftmark 和 rightmark，并将 leftmark 增加 1，将 rightmark 减少 1。如果 leftmark > pivot 和 rightmark > pivot，则只递减 rightmark。

②如果 leftmark < pivot 和 rightmark < pivot，则仅增加 leftmark。如果 leftmark < pivot 和 rightmark > pivot，则将 leftmark 增加 1，将 rightmark 减少 1。

③当 leftmark 和 rightmark 在同一个元素相遇时，将该元素与中心点互换。

（2）更新后的列表为 [2，6，4，13，50，39，28，15，20]，如图 13.13 所示。

①获取 13 左侧的元素，将 4 作为中心点，并以相同方式对其排序。

②左侧分区排序后，取右侧元素，并以 20 为中心点对其排序。

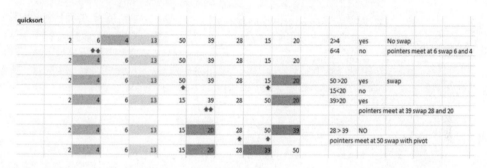

图 13.13

下面介绍快速排序的实现。

步骤 1：确定中心点。

（1）定义函数 find_pivot( )，使用 3 个参数——列表（input_list）、开始元素（first）和结束元素（last）的索引用于排序列表。

（2）使用 input_list.pivot = input_list[last] 将中心点设置为列表的最后一个值。

（3）将 left_pointer 设置为第一个。

（4）将 right_pointer 设置为 last − 1，因为最后一个元素是中心点。

（5）将 pivot_flag 设置为 True。

①如果 leftmark > pivot 和 rightmark < pivot，则交换 leftmark 和 rightmark，并将 leftmark 加 1，将 rightmark 减 1。

②如果 leftmark > pivot 和 rightmark > pivot，则只递减 rightmark。

③如果 leftmark < pivot 和 rightmark < pivot，则只增加 leftmark。

④如果 leftmark < pivot 和 rightmark > pivot，则将 leftmark 加 1，将 rightmark 减 1。

⑤当 leftmark 和 rightmark 在同一个元素处相遇时，将该元素与中心点交换。

⑥当 leftmark >= rightmark 时，将 pivot 的值与左指针处的元素交换，将 pivot_flag 设置为 False。

代码如下所示。

```python
def find_pivot(input_list, first, last):
    pivot = input_list[last]
    print("pivot =", pivot)
    left_pointer = first
    print("left pointer = ", left_pointer, " ", input_list[left_pointer])
    right_pointer = last - 1
    print("right pointer = ", right_pointer, " ", input_list[right_pointer])
    pivot_flag = True
    while pivot_flag:
        if input_list[left_pointer] > pivot:
            if input_list[right_pointer] < pivot:
                temp = input_list[right_pointer]
                input_list[right_pointer] = input_list[left_pointer]
                input_list[left_pointer] = temp
                right_pointer = right_pointer - 1
                left_pointer = left_pointer + 1
            else:
                right_pointer = right_pointer - 1
        else:
            left_pointer = left_pointer + 1
            right_pointer = right_pointer - 1
        if left_pointer >= right_pointer:
            temp = input_list[last]
            input_list[last] = input_list[left_pointer]
            input_list[left_pointer] = temp
            pivot_flag = False
    print(left_pointer)
    return left_pointer
```

步骤 2：定义 quickSort(input_list) 函数。该函数采用列表作为输入。

（1）确定排序的开始 point(0) 和结束 point(length_of_the_list − 1)。

（2）调用 qsHelper() 函数。

代码如下所示。

```python
def quickSort(input_list):
    first = 0
    last = len(input_list) - 1
    qsHelper(input_list, first, last)
```

步骤 3：定义 qsHelper() 函数。

qsHelper() 函数是递归函数，该函数调用 find_pivot() 函数，检查第一个索引和最后一个索引值。其中 leftmark 递增 1，rightmark 递减 1。因此，只要 leftmark（本示例中调用第一个参数）小于 rightmark（本示例中的最后一个参数），就会执行 while 循环，其中调用 qsHelper() 函数找到 pivot 的新值，创建 left 和 right 分区，并调用自身。代码如下所示。

```
def qsHelper(input_list,first,last):
    if first < last:
        partition = find_pivot(input_list,first,last)
        qsHelper(input_list,first,partition-1)
        qsHelper(input_list,partition+1,last)
```

代码：

```
def find_pivot(input_list, first,last):
    pivot = input_list[last]
    left_pointer = first
    right_pointer = last-1
    pivot_flag = True

    while pivot_flag:
        if input_list[left_pointer] > pivot:
            if input_list[right_pointer] < pivot:
                temp = input_list[right_pointer]
                input_list[right_pointer] = input_list[left_pointer]
                input_list[left_pointer] = temp
                right_pointer = right_pointer-1
                left_pointer = left_pointer+1
            else:
                right_pointer = right_pointer-1

        else:
            if input_list[right_pointer] < pivot:
                left_pointer = left_pointer+1
            else:
                left_pointer = left_pointer+1
                right_pointer = right_pointer-1

        if left_pointer >= right_pointer:
            temp = input_list[last]
            input_list[last] = input_list[left_pointer]
            input_list[left_pointer] = temp
            pivot_flag = False
    return left_pointer
def quickSort(input_list):
    first = 0
    last = len(input_list)-1
    qsHelper(input_list,first,last)
```

```
def qsHelper(input_list,first,last):
    if first < last:
        partition = find_pivot(input_list,first,last)
        qsHelper(input_list,first,partition-1)
        qsHelper(input_list,partition+1,last)
```

执行：

```
input_list =[15,39,4,20,50,6,28,2,13]
quickSort(input_list)
print(input_list)
```

输出：

```
[2, 4, 6, 13, 15, 20, 28, 39, 50]
```

本章介绍了几种非常重要的查找和排序算法，还介绍了如何使用Python对其进行实现。下一章将介绍如何使用Flask操作。

第 14 章

Flask框架入门

> 引言
>
> 本章简要介绍 Flask，这是一个非常流行的 Web 框架，广泛用于使用 Python 创建 Web 应用程序。

> 知识结构
>
> - 引言
> - 安装虚拟环境
> - 使用 Flask 开发"Hello World"应用
> - 调试 Flask 应用程序

> 目的
>
> 完成本章的学习后，读者应理解 Flask Web 框架，并且学会使用 Flask 创建简单网页。

14.1 引言

Flask 是非常流行的 Web 框架，广泛用于使用 Python 创建 Web 应用程序。由于其框架小，所以通常称之为微框架。Flask 的主要信息如下。

（1）Armin Ronacher 使用 Python 开发了 Flask。

（2）属于轻量级 Web 服务器网关接口（WSGI）Web 应用框架。

（3）免费用于商业用途。

（4）让初学者更容易进行 Web 开发，也适用于创建复杂的应用程序。

（5）理解 Flask 对 Python 程序开发人员大有裨益。

(6）提供创建 Web 应用程序的库、模块和工具。
(7）易于使用。
(8）具有内置的开发服务器和调试器。
(9）具有集成单元测试支持。
(10）使用 Jinja2 模板。
(11）支持安全 Cookie 客户端会话。
(12）100% 符合 WSGI 1.0 标准。
(13）基于 Unicode。
(14）支持文档应用。

Flask 的优势如下。
(1）可伸缩。
(2）开发简单。
(3）灵活。
(4）性能高。
(5）具有模块性。

Jinja2 是 Flask 使用的模板语言。Jinja 是 Python 的 Web 模板引擎，用于创建 HTML、HML 或其他标记格式，这些格式通过 HTTP 响应返回给用户。Python 代码也可用于创建 Web 应用程序。

14.2 安装虚拟环境

这里使用 pip 安装虚拟环境，代码如下所示。

```
pip install virtualenv
```

虚拟环境初衷是有一个在其中安装特定项目的软件包环境。假设在 Python 3.6 中创建了项目，如果直接通过邮件安装更新项目版本，则将破坏旧项目的某些模块。因此，每个项目用自己的软件包会更方便。

安装虚拟环境对于开发 Flask 应用程序非常重要。不建议部署 Flask 应用程序，尤其是使用 Python 创建的 Web 应用程序，因为可能导致很多"并发症"。在理想情况下，最好在干净的环境中工作，即能访问隔离的 Python 环境。仅将该环境用于使用 Flask 创建应用程序，而不作他用。这就要求编码前完成虚拟环境的创建。虚拟环境文件在与项目文件位于同一目录级别的文件夹中创建。

安装虚拟环境，必须使用 Python 3.3 版或更高版本（以及 2.7 版，但在本书中不使用 Python 2）以上内容。这里使用 Python 3.8.2 版本。

在开始前，请输入以下命令检查系统中已安装的软件包，如图 14.1 所示。

```
pip list
```

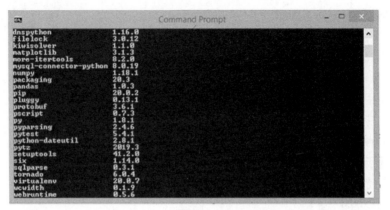

图 14.1

转到准备创建项目的目录。本示例在 C 盘中创建项目，所以转到 C 盘，如图 14.2 所示。

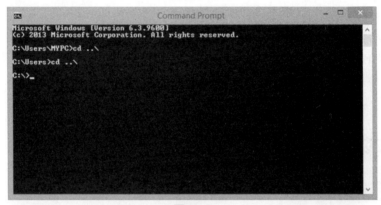

图 14.2

为项目创建文件夹，如图 14.3 所示。代码如下所示。

```
mkdir flask_project
```

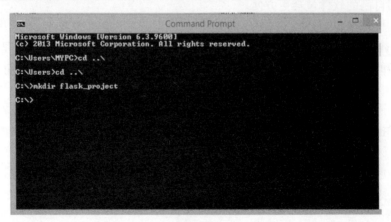

图 14.3

在 C 盘中创建新文件夹 "flask_project"，如图 14.4 所示。

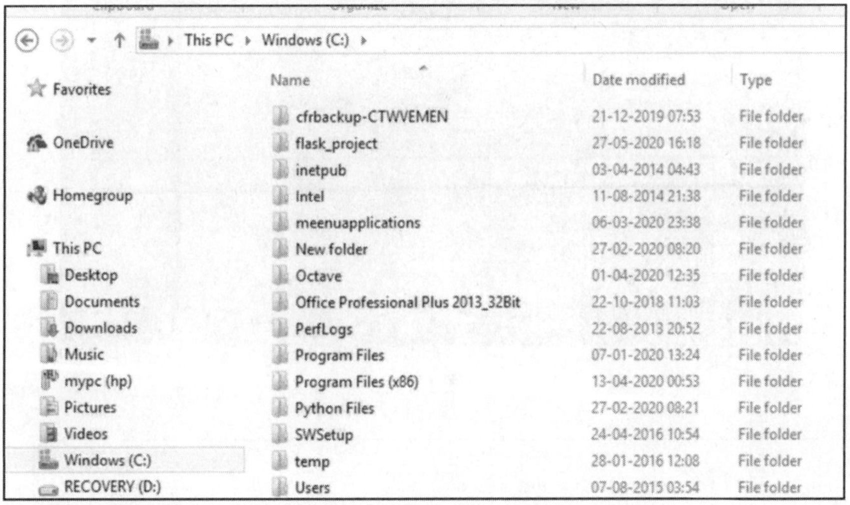

图 14.4

该地址是工作地点，如图 14.5 所示。

```
cd flask_project
```

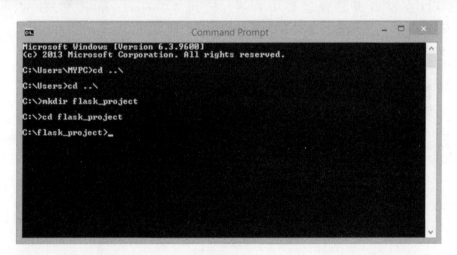

图 14.5

要创建新环境，请输入以下命令（见图 14.6）：

```
python -m venv virtual_environment_name
```

输入 "dir"，会看到新环境 flvirenv 已在 "C:\flask_project" 文件夹中创建，如图 14.7 所示。

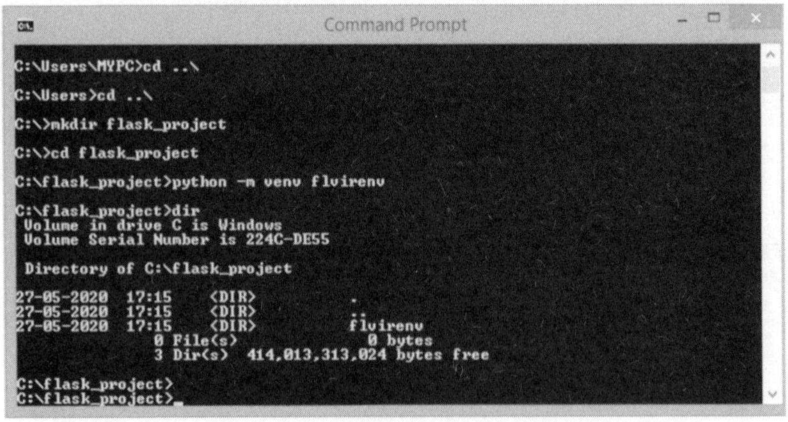

图 14.6

图 14.7

也可以在窗口资源管理器中查看。转到"flvirenv/Scripts"文件夹，看到"activate.bat"文件，如图 14.8 所示，然后激活虚拟环境。

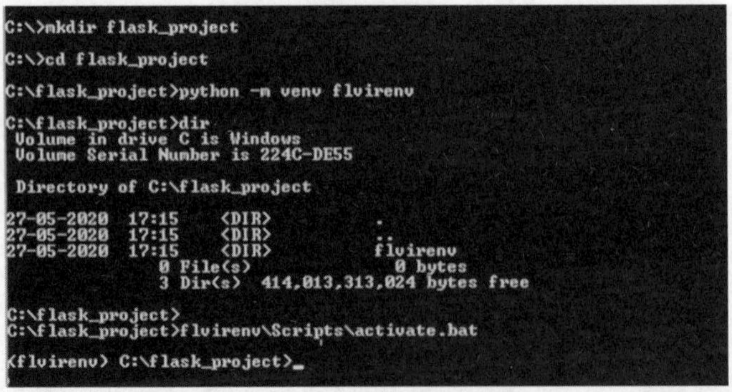

图 14.8

激活虚拟环境后，可在提示符左侧的括号中看到其名称。环境中使用的 Python 版本与用于创建的版本相同。现在输入"pip list"，以确定环境中安装的所有内容。在此环境中，安装处理项目所需的所有内容。

通过以下命令安装 Flask（见图 14.9）：

```
pip install flask
```

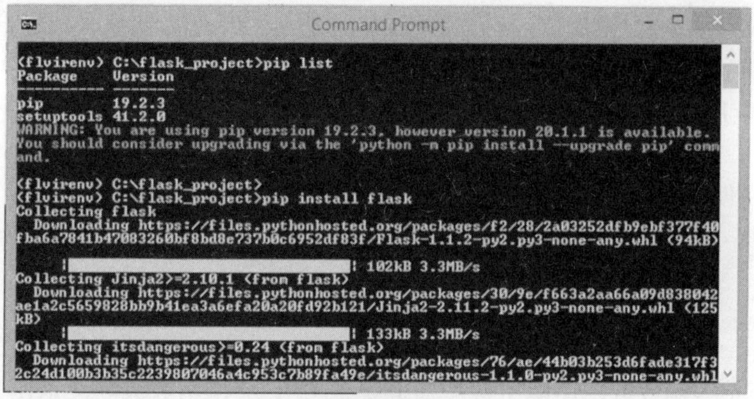

图 14.9

再次输入"pip list",确定"flvirenv"文件夹中已安装的内容,如图 14.10 所示。

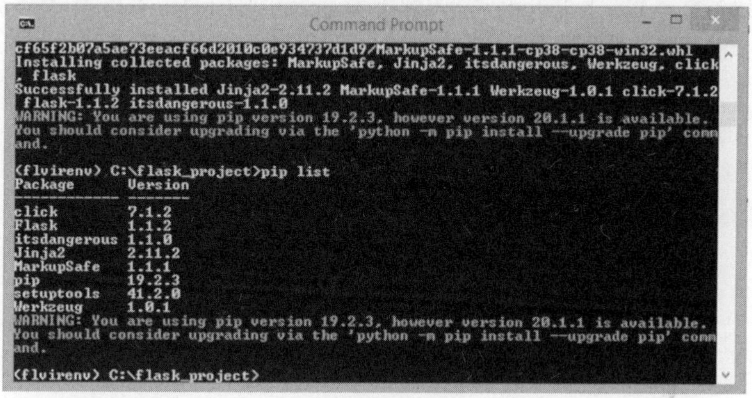

图 14.10

14.3 使用 Flask 开发"Hello World"应用

下面在 Flask 框架中创建第一个应用。

步骤 1:导入 Flask 类。代码如下所示。

```
from flask import Flask
```

Flask 类的实例是 WSGI 应用程序,也是 Flask 框架中构建应用程序的中心对象。WSGI 代表 Web 服务器网关接口,是 Python 增强提案(PEP)3333 中定义的 Python 标准 Web 服务器接口。WSGI 是简单调用约定,允许 Python Web 应用程序与 Web 服务器进行交互。程序开发人员只需专注开发应用程序,Flask 会处理所有必要过程的实现。前提是导入 Flask 类。

步骤 2:创建 Flask 类的实例。

Flask 实例通常在主模块或"init.py"文件的包中创建。代码如下所示。

```
app = Flask( __name__ )
```

其中，参数"__name__"用于查找文件系统上的资源。如果计划创建单个模块（如正在处理的应用程序），则是正确的值。但是，如果使用包，则必须提供包名称作为参数。

步骤 3：使用 route( ) 装饰器。代码如下所示。

```
@app.route('/hello')
def helloIndex():
return 'Hello World!'
```

调用 route( ) 函数可以使应用程序知道哪个 URL 与该函数关联。该函数由 URL 触发。基本上对根 URL('/hello') 的所有请求都将定向到下一行中定义的函数 helloIndex( )。

@ app. route 修饰器会包装 helloIndex( ) 函数，将 URL（'/hello'）的请求路由转到特定视图。函数名称不限。helloIndex( ) 函数不带参数并返回简单字符串。

步骤 4：输入命令运行应用程序。代码如下所示。

```
if __name__ == '__main__':
    app.run(port =5000,debug =True)
```

整体代码如图 14.11 所示。

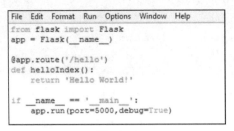

图 14.11

现在，按照以下步骤运行该应用程序。

步骤 1：将文件保存在项目文件夹中。

将文件的名称"hello. py"保存在"flask_project"文件夹中，如图 14.12 所示。

图 14.12

步骤2：激活虚拟环境，如图 14.13 所示。

图 14.13

步骤3：输入以下命令，如图 14.14 所示。

```
python hello.py
```

图 14.14

步骤4：在浏览器中打开应用程序。

在命令提示符下，看到应用程序正在 http://127.0.0.1:5000 上运行，代码中已将 app.route 设置为 '/hello'。因此，查看应用程序，必须输入 "http://127.0.0.1:5000/hello"，如图 14.15 所示。

图 14.15

现在尝试使用动态路由，旨在从动态 URL 中读取人员姓名并欢迎他/她。在"hello.py"文件中添加另一个函数，如图 14.16 所示。

```
from flask import Flask
app = Flask(__name__)

@app.route('/hello')
def helloIndex():
    return 'Hello World!'

@app.route('/welcome/<name>')
def welcomeUser(name):
    return '<h1> {} We Welcome you to our Page!!'.format(name)

if __name__ == '__main__':
    app.run(port=5000,debug=True)
```

图 14.16

输出结果如图 14.17 所示。

图 14.17

此例用途如下所示。

```
@app.route('/welcome/<name>')
def welcomeUser(name):
    return'<h1> {} We Welcome you to our Page!! '.format(name)
```

app.route 装饰器将路由设置为'/welcome/<name>'。其中，name 是动态更改参数。定义 welcomeUser()函数，括号中也有相同参数。

输入"http://127.0.0.1/welcome/Michael"时，Flask 将自动映射，<name>在本例中是将 Michael 作为 name 参数传递给函数 welcomeUser()。该函数将姓名添加到欢迎消息中并将其显示在浏览器上。这里没有使用浏览器，而是使用 html 标记<h1>将消息显示为标题。

14.4 调试 Flask 应用程序

因为应用程序中不可避免会出现漏洞和缺陷，所以为了使程序开发更简单，需要将调试模式设置为 True 或 False。运行应用程序时，通过输入"app.run（debug = True）"来设置调试模式。从图 14.18 中不难看出，其中代码的第 2 行存在错误，但通过设置 debug = True 对其进行识别。

```
from flask import Flask
app = Flask(__name__)

@app.route('/hello')
def helloIndex():
    return 'Hello World!'

@app.route('/welcome/<name>')
def welcomeUser(name):
    return '<h1> {} We Welcome you to our Page!!'.format(name)

if __name__ == '__main__':
    app.run(port=5000,debug=True)
```

图 14.18

尝试运行代码时，发生图 14.19 所示情况。

图 14.19

小结

  Flask 是用 Python 编写的 Web 应用程序框架。使用它更容易创建和实现 Web 应用程序。读者学习完使用 Flask 显示简单网页的基础知识，接下来学习本书的附录。

附 录

1. Big-O 表示法简述

Big-O 表示法用于分析在流程中数据增加的情况下，算法将如何执行，简单来说，它是对算法效率的简化分析。

由于算法是软件编程的重要组成部分，所以必须理解算法运行所需的时间。只有这样，才能比较两种算法，好的程序开发人员在规划编码过程时总是会考虑时间复杂度，这有助于确定算法的运行时间。

2. Big-O 表示法的重要性

Big-O 表示法描述算法运行时相对输入规模增加的速度，会对算法扩展有清晰的认识，也能理解目前项目算法的最坏情况复杂度。可以使用 Big-O 表示法比较两种算法，并确定哪种算法更好。Big-O 表示法很重要，原因如下。

（1）提供输入大小 n 的算法复杂度。

（2）只考虑算法涉及步骤。

（3）可用于分析时间和空间。

算法效率用最佳平均值或最坏情况衡量，但 Big-O 表示法适用于最坏情况。

以不同方式执行任务，因此同一任务有不同算法，具有不同复杂度和可伸缩性。

Big-O 表示法用于比较各种类型输入大小的运行时间。重点仅关注输入大小对运行时间的影响。随着 n 值的增大，只关心它如何影响运行时间。如果直接测量运行时间，那么测量单位将是时间单位，如秒、微秒等。但是，在本示例中，n 代表输入大小，O 代表阶。因此，O(1) 代表 1 的阶；O(n) 代表 n 的阶；O( ) 代表输入大小的平方阶。

现在，假设有两个函数。

1）恒定复杂度 O(1)

输入任意值，常量任务的运行时间都不会变化。例如：

```
>>> x = 5 + 7
>>> x
12
>>>
```

语句 x = 5 +7 不依赖数据输入大小。这称为 O(1)（恒 1）。

示例：

假设所有常量时间都有一系列步骤，代码如下所示。

```
>>> a =(4 -6) + 9
>>> b =(5 * 8) +8
>>>print(a * b)
336
>>>
```

按以下步骤计算 Big-O。
$$总时间复杂度 = O(1) + O(1) + O(1)$$
$$= 3O(1)$$

在计算 Big-O 时，忽略常量，因为数据大小增加，常量值无关紧要，所以 Big-O 是 $O(1)$。

2）线性复杂度 $O(n)$

在线性复杂度的情况下，运行时间取决于输入值。

假设要打印 5 个表，查看下面的代码。

```
>>> for i in range(0,n):
 print("\n 5 x ",i," = ",5 * i)
```

打印行数取决于 n。n = 10，只打印 10 行，但 n = 1 000，执行 for 循环将花费更多时间。打印语句为 $O(1)$。因此，代码块是 $n * O(1)$，即 $O(n)$。分析以下代码行。

```
>>>j = 0                           ------(1)
>>> for i in range(0,n):           ------(2)
    print("\n 5 x ",i," = ",5 * i)
```

语句（1）为 $O(1)$；语句块（2）为 $O(n)$。
$$总时间复杂度 = O(1) + O(n)$$

因为 $O(1)$ 是低阶项，可以将其删除，当 n 值变为恒（1）时，运行时间实际上取决于 for 循环。因此，上述代码的 Big-O 是 $O(n)$。

3）二阶复杂度 $O()$

二阶复杂度所花费时间取决于输入值大小。嵌套循环就是这种情况。请看下面的代码。

```
>>> for i in range (0,n):
    for j in range(0,n):
        print("I am in ", j," loop of i = ", i,".")
```

在此代码中，print 语句执行了一段时间。

4）对数复杂度

对数复杂度表明，在最坏情况下，算法必须执行 logn 步骤。为理解这一点，首先理解对数概念。

对数是幂的逆过程。

这里 2 是基数，3 是指数。因此，以 2 为基数 8 的对数等于 3。同样，以 10 为基数的 100 000 的对数等于 5。

计算机使用二进制数，因此，在编程和 Big – O 中通常使用基数为 2 的对数。

观察下面的结果。

(1) $\log_2^n$，n = 2；

(2) $\log_2^n$，n = 4；

(3) $\log_2^n$，n = 8；

(4) $\log_2^n$，n = 16；

(5) $\log_2^n$，n = 32。

如果 n = 2，步数 = 1。

如果 n = 4，步数 = 2。

如果 n = 8，步数 = 3。

因此，随着数据加倍，步数增加 1。相比数据大小的增长，步数增长缓慢。

编程中对数复杂度的最佳示例是二叉搜索树。通过第 12 章读者可了解更多相关信息。

示例：

```
i = j = k =0
for i in range(n/2,n):
    for j in range(2,n):
        k =k+n/2
                    j = j*2
```

第 1 个 for 循环的时间复杂度为 O(n/2)。

第 2 个 for 循环的时间复杂度为 O(logn)，因为 j 是自身加倍，直到它小于 n，所以

$$总时间复杂度 = O(n/2) * O(logn)$$
$$= O(nlogn)$$

3. Big – O 符号和名称

(1) 常量——O(1)。

(2) 对数——O(logn)。

(3) 线性——O(n)。

(4) 对数线性——O(nlogn)。

(5) 平方、立方、指数——O( )。

计算机科学中最坏情况的时间复杂度意味着执行程序时消耗时间方面最坏的情况。这是算法花费的最长运行时间。算法效率通过查看最坏情况的时间复杂度的增长顺序来比较。

Big – O 表示法属于渐近表示法。

理解程序运行速度非常重要。不同计算机具有不同硬件功能。不考虑确切运行时

间，因为结果可能会因计算机的硬件、处理器的速度，以及在后台运行的其他处理器而异。更重要的是要理解算法的性能如何随着输入数据的增加而受到影响。

为了解决该问题，人们提出了渐近表示法的概念。这为测量算法速度和效率提供了一种通用方法。对于处理大输入的应用程序，随着输入大小的增加，人们会对算法效果更感兴趣。Big – O 表示法是最重要的渐近表示法之一。

4. 时间和空间的复杂度

时间复杂度给出解决问题所涉及步骤数量方面问题的想法。复杂度顺序（升序）如下：

$$O(1) \ < \ O(\log n) \ < \ O(n) \ < \ O(n \log n) \ < \ O(\ )$$

与时间不同，内存可以重用，人们对计算速度更感兴趣。这是人们经常讨论时间复杂度而不是空间复杂度的主要原因之一。然而，空间复杂度永远不会被忽视。空间复杂度决定整个运行程序所需的内存量。以下情况需要内存。

（1）指令空间。

（2）包含常量、变量、结构化变量、动态更改区域等的知识空间。

（3）执行。

顾名思义，空间复杂度描述内存空间大小。如果输入的大小 n 增加，则为必需。这里考虑最坏情况。现在请观察以下代码：

```
>>> x = 23              (1)
>>> y = 34              (2)
>>> sum = x + y         (3)
>>> sum
57
>>>
```

在这个示例中，需要空间存储 3 个变量：（1）中的 x、（2）中的 y 和（3）中的 sum。但是，该情况不会改变，并且对 3 个变量的要求恒定，因此空间复杂度为 O(1)。

现在，观察以下代码：

```
word = input("enter a word : ")
word_list = []
for element in word:
    print(element)
    word_list.append(element)
    print(word_list)
```

word_list 对象的大小随 n 而增大。因此，在本例中，空间复杂度为 O(n)。

如果假设 function1( ) 函数有 3 个变量，并且 function1( ) 函数调用另一个有 6 个变量的 function2( ) 函数，那么临时工作区的总体要求为 9 个单位。即使 function1( ) 函数调用 function2( ) 函数 10 次，工作区要求也保持不变。

请观察以下代码：

```
n = int(input("provide a number : "))
statement = "Happy birthday"
for i in range(0,n):
    print(i+1,". ",statement)
```

上述代码的空间复杂度为 O(1)，因为空间要求仅用于存储整数变量 n 和字符串语句的值。即使 n 的大小增加，空间要求也保持不变，因此空间复杂度为 O(1)。

现在，分析以下代码的时间复杂度和空间复杂度：

```
a = b = 0
for i in range(n):
    a = a + i
for j in range(m):
    b = b + j
```

（1）时间复杂度。

第 1 个循环的时间复杂度为 O(n)，第 2 个循环的时间复杂度为 O(m)。

$$总时间复杂度 = O(n) + O(m) = O(n+m)$$

（2）空间复杂度。

空间复杂度为 O(1)。

读书笔记

读书笔记